Lecture Notes in Physics

New Series m: Monographs

Managing Editor

W. Beiglböck
Assisted by Mrs. Sabine Landgraf
c/o Springer-Verlag, Physics Editorial Department II
Tiergartenstrasse 17, D-69121 Heidelberg, Germany

The Editorial Policy for Monographs

The series Lecture Notes in Physics reports new developments in physical research and teaching - quickly, informally, and at a high level. The type of material considered for publication in the New Series m includes monographs presenting original research or new angles in a classical field. The timeliness of a manuscript is more important than its form, which may be preliminary or tentative. Manuscripts should be reasonably self-contained. They will often present not only results of the author(s) but also related work by other people and will provide sufficient motivation, examples, and applications.

The manuscripts or a detailed description thereof should be submitted either to one of the series editors or to the managing editor. The proposal is then carefully refereed. A final decision concerning publication can often only be made on the basis of the complete manuscript, but otherwise the editors will try to make a preliminary decision as definite as they can on the basis of the available information.

Manuscripts should be no less than 100 and preferably no more than 400 pages in length. Final manuscripts should preferably be in English, or possibly in French or German. They should include a table of contents and an informative introduction accessible also to readers not particularly familiar with the topic treated. Authors are free to use the material in other publications. However, if extensive use is made elsewhere, the publisher should be informed. Authors receive jointly 50 complimentary copies of their book. They are entitled to purchase further copies of their book at a reduced rate. As a rule no reprints of individual contributions can be supplied. No royalty is paid on Lecture Notes in Physics volumes. Commitment to publish is made by letter of interest rather than by signing a formal contract. Springer-Verlag secures the copyright for each volume.

The Production Process

The books are hardbound, and quality paper appropriate to the needs of the author(s) is used. Publication time is about ten weeks. More than twenty years of experience guarantee authors the best possible service. To reach the goal of rapid publication at a low price the technique of photographic reproduction from a camera-ready manuscript was chosen. This process shifts the main responsibility for the technical quality considerably from the publisher to the author. We therefore urge all authors to observe very carefully our guidelines for the preparation of camera-ready manuscripts, which we will supply on request. This applies especially to the quality of figures and halftones submitted for publication. Figures should be submitted as originals or glossy prints, as very often Xerox copies are not suitable for reproduction. For the same reason, any writing within figures should not be smaller than 2.5 mm. It might be useful to look at some of the volumes already published or, especially if some atypical text is planned, to write to the Physics Editorial Department of Springer-Verlag direct. This avoids mistakes and time-consuming correspondence during the production period.

As a special service, we offer free of charge $\LaTeX$ and $\TeX$ macro packages to format the text according to Springer-Verlag's quality requirements. We strongly recommend authors to make use of this offer, as the result will be a book of considerably improved technical quality.

Manuscripts not meeting the technical standard of the series will have to be returned for improvement.

For further information please contact Springer-Verlag, Physics Editorial Department II, Tiergartenstrasse 17, D-69121 Heidelberg, Germany.

Johnny T. Ottesen

Infinite Dimensional Groups and Algebras in Quantum Physics

 Springer

Author

Johnny T. Ottesen
Department of Studies in Mathematics and Physics
and their Functions in Education, Research and Applications
Roskilde University, P. O. Box 260
DK-4000 Roskilde, Denmark

ISBN 978-3-662-14053-6   ISBN 978-3-540-49141-5 (eBook)
DOI 10.1007/978-3-540-49141-5

CIP data applied for.

Typesetting: Camera-ready by the author
SPIN: 10127317   55/3142-543210 - Printed on acid-free paper

To Tove, Elisabeth and Katrine

Preface

The idea of writing this book appeared when I was working on some problems related to representations of physically relevant infinite dimensional groups of operators on physically relevant Hilbert spaces. The considerations were local, reducing the subject to dealing with representations of infinite-dimensional Lie algebras associated with the associated groups.

There is a large number of specialized articles and books on parts of this subject, but to our suprise only a few represent the point of view given in this book. Moreover, none of the written material was self-contained.

At present, the subject has not reached its final form and active research is still being undertaken. I present this subject of growing importance in a unified manner and by a fairly simple approach. I present a route by which students can absorb and understand the subject, only assuming that the reader is familliar with functional analysis, especially bounded and unbounded operators on Hilbert spaces. Moreover, I assume a little basic knowledge of C^*-algebras, Lie algebras, Lie groups, and manifolds – at least the definitions. The contents are presented in detail in the introduction in Chap. 1.

The manuscript of this book has been succesfully used by some advanced graduate students at Aarhus University, Denmark, in their "A-exame". I thank them for comments.

I want to aknowledge the "Department of Mathematics and Physics and their Functions in Education, Research and Applications" at Roskilde University Centre, Denmark, for giving us the opportunity to write this book. Especially, I thank Anders Madsen, Gestur Olufson, Woicheck Slovikowski, and Bent Ørsted for reading the manuscript and for making numerous suggestions and corrections. In particular, I want to thank Lars-Erik Lundberg for fruitful discussions and for providing some ideas and results.

Roskilde University Centre
January 16, 1995

Johnny T. Ottesen

Contents

Chapter 1

Introduction

Recently there has been a renewed interest in representations related to the study of Kac-Moody algebras, loop algebras, the Virasoro algebra and the associated groups (see for example [P-S], [K-R], [G-W] and [Ar]). These groups appear naturally in quantum theory with infinitely many degrees of freedom, i.e. in canonical quantum field theory, string theory, statistical quantum physics and soliton theory. The renewed interest comes partly from the many fruitful applications in physics and partly, or rather basically, from the mathematical success in dealing with infinite dimensional Lie algebras.

One of the problems in dealing with infinite dimensional Lie algebras is that many of the essential methods known from finite dimensional Lie algebras cannot be suitably generalized. In particular, such operations as exponentiating elements of the Lie algebras are not always possible, even locally.

The loop group is the group of smooth mappings from the unit circle S^1 to a Lie group G, and it is denoted LG (sometimes $S^1(G)$ or $Map(S^1, G)$). If the Lie algebra of the Lie group G is $\mathfrak{g}$, then the Lie algebra of LG is $S^1\mathfrak{g}$ (sometimes $l\mathfrak{g}$ or $Map(S^1,\mathfrak{g})$). The Lie algebra $S^1\mathfrak{g}$ is called the loop algebra of $\mathfrak{g}$. Hereby the loop algebra gives information on the loop group. The central extension of the loop algebra $S^1\mathfrak{g}$ is denoted the affine Kac-Moody algebra associated with the Lie algebra $\mathfrak{g}$. Sometimes one needs a further extension of an affine Kac-Moody algebra, this is also called an affine Kac-Moody algebra or just an affine algebra, we prefer the last nomenclature.

From a mathematical point of view, the Virasoro algebra is a relatively well-behaved infinite dimensional Lie algebra, and its representations are well understood. The Virasoro algebra is the central extension of the complexification of the smooth real vector fields on the unit circle S^1, hence it is a central extension of the complexification of the Lie algebra of the diffeomorphism group. The exponential mapping from the Lie algebra of real smooth vector fields on the unit circle to the diffeomorphism group is neither locally one-to-one nor onto. It turns out that the diffeomorphism group on the unit circle $Diff(S^1)$ acts as a group of automorphisms on any loop group and that the orientation preserving subgroup $Diff^+(S^1)$ acts projectively on all the known representations of loop groups. Hence, it is natural to study the Virasoro algebra in connection with the loop algebras.

The loop algebras and the Virasoro algebra combine in the Sugawara construction, which defines the generators of a representation of the Virasoro algebra as quadratic expressions in the basis of any representation of the affine algebra corresponding to the mentioned loop algebra. This turns out to be very important in conformally invariant statistical physics.

In a standard approach to quantum field theory the states of the physical system are vectors in a Fock Hilbert space. The precise structure of the Fock Hilbert space depends on the type of statistics obeyed by the particles it describes. In the case of bosons one uses the symmetric tensor algebra, $\mathcal{F}_\vee(\mathcal{H})$, modelled over a "one-particle Hilbert space" $\mathcal{H}$, to describe the physical many particle states, hence it is sometimes called the boson Fock Hilbert space. When we deal with fermions, the physical many particle state is described by the Hilbert space completion, $\mathcal{F}_\wedge(\mathcal{H})$, of the exterior algebra over a one-particle Hilbert space $\mathcal{H}$; it is sometimes denoted the fermion Hilbert space. The basic one-particle Hilbert space will, in our considerations, always be separable and infinitely dimensional.

The symmetry of the Fock Hilbert spaces (or, in a physical language, of the particles) are reflected in the commutation relations. In the fermion case, the abstract creation and annihilation operators, which generate a C^*-algebra, parameterized by vectors in a Hilbert space, fulfil the canonical anti-commutation relations, CAR. This C^*-algebra is called the CAR-algebra or sometimes the fermion algebra. It has a very useful realization on the anti-symmetric Fock Hilbert space, called the Fock representation of the CAR-algebra. In the case of bosons, the abstract creation and annihilation operators generate (only) a *-algebra parameterized by vectors in a Hilbert space and fulfil the canonical commutation relations, CCR. This *-algebra is sometimes denoted the boson algebra. However, the boson algebra gives rise to the Weyl operators, which are unitaries constructed from the creation and annihilation operators, and they generate a C^*-algebra, called the CCR-algebra. Similar to the anti-symmetric case, there are very useful realizations of the boson algebra and the

CCR-algebra in the symmetric Fock Hilbert space, called the Fock representation of the boson algebra and the *CCR*-algebra, respectively. It turns out that both the fermion and the boson algebras are unique up to an isomorphism, whereby we can (and will) benefit from the Fock representations of these in the study of certain automorphism groups.

The automorphism groups are, in proper circumstances, unitarily implemented in the Fock representation. These circumstances are closely related to the restricted orthogonal group and the restricted symplectic group, in the case of the fermion algebra and the boson algebra, respectively. The unitary implementers can be constructed explicitly as projective representations of the restricted orthogonal group and the restricted symplectic group, respectively. It turns out that the loop group, LS^1, can be realized as a subgroup of the restricted orthogonal group and that the orientation preserving diffeomorphism group can be realized both as a subgroup of the restricted symplectic group and as a subgroup of the restricted orthogonal group. This will be used to construct particular representations of these groups on a Lie algebra level.

This book is an attempt to give a self-contained review of the spin representation of the infinite dimensional restricted orthogonal group and of the metaplectic representation of the infinite dimensional restricted symplectic group, and their relations to the fermion algebra and the boson algebra, respectively. Moreover, we clarify the connection of these topics with the theory of loop algebras, Kac-Moody algebras and the Virasoro algebra, which we also discuss in a self-contained manner. Especially, we apply the representations to the particular loop group LS^1 and the diffeomorphism group $Diff^+(S^1)$. Furthermore, we construct various Fock Hilbert space representations of the Virasoro algebra.

Our major tool in dealing with these topics is functional analysis, working mostly on a Lie algebra level. We emphasize that we only consider separable infinite dimensional Hilbert spaces, hence the groups and algebras appearing will also be of infinite dimension.

We will now turn to some physical motivation. The subjects of this book are strongly related to quantum field theory, a widely bifucated and very theoretically branch of physics (see [F-M-S], [Ka], [Ka], [V-Z]; a more kaleidoscopic impression can be found in [F-H-J]). We will briefly explain how the Virasoro algebra (and $Vect(S^1)$), the diffeomorphism groups and the loop groups enter into physics.

Somehow physics is the science concerned with symmetries of nature. Since symmetries of a given physical system form a group, representation theory is of interest. In the study of crystals the relevant groups are the finite or discrete ones, for example, certain translations with the distance given by the lattice length. In the study of atomic spectra it is the finite continuous groups, such as the group of rotations in three dimensions, $SO(3)$, which are used. Since space and time are (supposed) continuous, one expects continuous (Lie) groups to enter physics too. In quantum physics one usually expresses the symmetries naturally through a Lie group of Hermitian operators acting on an a priori Hilbert space, leaving the physical system (i.e. the Hamiltonian) invariant. Thereby one is led to consider the Lie algebra of generators of the Lie group. For example, the angular momenta generate $SO(3)$, hence they define an interesting Lie algebra. The simplest class of infinite dimensional Lie algebras, naturally generalizing the properties of finite dimensional Lie algebras, are the affine Kac-Moody algebras, and there is always associated with each affine Kac-Moody algebra a Virasoro algebra. During the last twenty years an increasing number of physical models and theories have been presented using one or both of the above-mentioned infinite dimensional algebras in an essential way. We will discuss some examples from physics where the Virasoro algebra enters.

The first example appears in two-dimensional conformally invariant quantum physics, see for example [B-P-Z 1], [F-M-S], [Ka] or [L-T] where more details can be found. Introduce complex coordinates z and $\bar{z}$ in the two-dimensional Euclidian plane, $z = x + iy$ and $\bar{z} = x - iy$. The infinitesimal conformal transformations of the plane is given by the infinite number of generators $L_n = z^{n+1}\partial_z$ and $\bar{L}_n = \bar{z}^{n+1}\partial_{\bar{z}}$, where $\partial_z = \frac{d}{dz}$ and $\partial_{\bar{z}} = \frac{d}{d\bar{z}}$. In conformally invariant systems with periodic boundary conditions the generators L_n and $\bar{L}_n$ appear as coefficients of the Laurent expansion of the component of the energy-momentum tensor (see below). Symmetry and tracelessness of the energy-momentum tensor implies that it can be described by two independent components. In the $(z, \bar{z})$-coordinates we may choose $T = T'_{00} + T'_{01}$ and $\bar{T} = T'_{00} - T'_{01}$ as independent varibles, where the primes indicate that it is in the old coordinates. Then the conservation (or continuity) equation implies that $\partial_{\bar{z}}T = \partial_z\bar{T} = 0$, hence $T = T(z)$ and $\bar{T} = \bar{T}(\bar{z})$ are both holomorphic in z and $\bar{z}$, respectively. The energy-momentum tensor is then described in terms of the holomorphic components

$$T = T(z) \qquad \text{and} \qquad \overline{T} = \overline{T}(\overline{z})$$

which we may Laurent expand as

$$T(z) = \sum_{n \in \mathbf{Z}} z^{-n-2} L_n \qquad \text{and} \qquad \overline{T}(\overline{z}) = \sum_{n \in \mathbf{Z}} \overline{z}^{-n-2} \overline{L}_n$$

where

$$L_n = \frac{1}{2\pi i} \oint_{z=0} T(z) z^{n+1} \, dz$$

and

$$\overline{L}_n = \frac{1}{2\pi i} \oint_{\overline{z}=0} \overline{T}(\overline{z}) \overline{z}^{n+1} \, d\overline{z}$$

The self-adjointness of T and $\overline{T}$ implies that $L_n^* = L_{-n}$ and $\overline{L}_n^* = \overline{L}_{-n}$ and it follows that each of the family of generators fulfils the commutation relations of the Virasoro algebra. In quantum field theory one usually describe the conformally invariant systems in terms of the correlation functions of the primary fields ϕ (i.e. fields fulfilling the particular transformation law, $\phi(z) \rightarrow f(\phi(z))^h \phi(f(z))$, under conformal transformations f in contrast to secondary fields). These fields satisfy $[L_n, \phi] = z^{n+1} \partial_z \phi + h(n+1) z^n \phi$ and the same with barred symbols $\overline{z}, \overline{L}_n$ and $\overline{h}$ instead of z, L_n and h, respectively, where h and $\overline{h}$ are the (energy) eigenvalues of L_0 and $\overline{L}_0$, respectively, corresponding to the eigenvector $\phi(0)\Omega$ (here Ω denotes the vacuum vector). Thus, it can be shown that the two-point function, describing the field-correlation between point z_1 and z_2, becomes

$$\langle \Omega, \phi(z_1)\phi(z_2)\Omega \rangle = r^{-2(h+\overline{h})} e^{-2i\theta(h-\overline{h})} \langle \Omega, \phi(0)\phi(0)\Omega \rangle$$

where $z_2 - z_1 = r e^{i\theta}$. The first factor $r^{-2(h+\overline{h})}$ is a consequence of demanding rescaling (dilation) invariance and the second factor of demanding rotation invariance. The $h + \overline{h}$ is called the scaling dimension and is clearly related to the energy, and $h - \overline{h}$ is referred to as the conformal spin of the field ϕ. Hereby, one reads the power law of the spatial separation, i.e. that the correlation function decays with the separation to the power $-2(h + \overline{h})$.

The second example is a continuation of the first. In statistical physics, one often uses lattice models as follows. Consider a two-dimensional lattice with a spin variable at each vertex site, for

example the Ising model (which is used to decribe such physical phe-
nomena as ferromagnetism of materials such as iron; see for example
[F-H-J]). In a tractable approximation one supposes only interaction
between nearest neighbours. At a certain critical point, for example
a temperature point, the physical system can make a second-order
phase transition (i.e. the entropy is continuous as a function of temperature but its derivative is discontinuous), whereby it changes its
physical state. It turns out that the lattice spacing becomes irrele-
vant, giving a scale invariant theory. In the conformally invariant
case the theory, or model, is controlled by the Virasoro algebra,
whose representation theory determines the critical exponents of the
transition (see [B-P-Z 1]). As mentioned above, the critical expo-
nent specifies certain power law behaviours in spatial seperations,
which are measurable in the laboratory.

The final example concerns how the Virasoro algebra enters the
theory of strings through the diffeomorphism group (following [Se,G.,
p.336]). As a crude oversimplification one can say that one wants
to describe unparameterized strings but finds it more convenient to
describe parameterized strings. The group of diffeomorphisms act on
the Hilbert space of states of parameterized strings by changing the
parameterization. Following G. Segal [Se,G., p.336] a string is a one
dimensional object, a mathematical curve on a manifold, i.e. it is the
image of a smooth map $(x : [0,1] \to \mathbf{R}^3)$. When it moves it sweeps
out a "world-surface" $(x : [0,1] \times \mathbf{R} \to \mathbf{R}^4)$ in Minkowski space-
time. The points of the string are regarded as indistinguishable,
so the parameterization has no physical significance (see also [Mi,
Chap. 9]).

We will now outline the contents of this book systematically.
Each section begins with some comments. Some are historical re-
marks, others are an attempt to explain the connection between the
particular section and the rest of the contents. Moreover, we provide
a list of the main references related to the particular section.

Chapter 2 is a survey on the spin representation of the infinite
dimensional orthogonal group in the anti-symmetric Fock Hilbert
space. Most of it is well known, but we have rewritten the material in
a collected form. The results of section 2.6 do not, to our knowledge,
appear anywhere else.

In *section* 2.1 we introduce the Fock Hilbert spaces together with
some of their important subspaces. The explicit constructions will
be used extensively in this and the following chapters.

In *section* 2.2 we describe, up to a *-isomorphism, the unique C^*-algebra called the CAR-algebra over a Hilbert space and its Fock representation. Here CAR stands for canonical anti-commutation relations. Sometimes the CAR-algebra is also called the fermion algebra. The CAR-algebra has an equivalent formulation in terms of the Clifford algebra, which we also discuss. Since the CAR-algebra, and therefore the Clifford algebra, too, are "unique", we analyse them in their Fock representations. Moreover, we ask the question: for which orthogonal operators is the automorphism, $\pi(f) \to \pi(T^{-1}f)$, of the Clifford algebra, unitarily implementable in the Fock representation?

In *section* 2.3 we use the explicit construction of the anti-symmetric Fock Hilbert space and the Fock representation of the CAR-algebra to discuss the construction of second quantization (on the anti-symmetric Fock Hilbert space). This second quantization is essential for our construction of the spin representation, treated in the next section.

In *section* 2.4 we construct the so-called spin representation of the infinite dimensional restricted orthogonal group, on a Lie algebra level, by use of the second quantization in the anti-symmetric Fock Hilbert space and some generalizations of this idea. This discussion is closely related to the question stated in section 2.2, since the automorphism, $\pi(f) \to \pi(T^{-1}f)$, of the Clifford algebra, is unitarily implementable in the Fock representation if and only if the orthogonal transformation T belongs to the restricted orthogonal group $\mathcal{O}_2(\mathcal{H})$. In fact, this answer was given by Shale and Stinespring in 1965. However, our methods are quite different from theirs, in that we construct the representation explicitly on the subgroup consisting of one-parameter groups in a neighbourhood of the identity in $\mathcal{O}_2(\mathcal{H})$. We finally end this section by giving an explicit formula for the vacuum functional. Beyond giving an explicit projective representation of the restricted orthogonal group, this section will be used, in Chap. 5 on applications, to make an explicit representation of the orientation preserving diffeomorphism group, $Diff^+(S^1)$, on the unit circle, on a Lie algebra level, by realizing $Diff^+(S^1)$ as a subgroup of the (enlarged) restricted orthogonal group on a particular Hilbert space.

In *section* 2.5 we consider the restriction of the spin representation to the infinite dimensional restricted unitary group, viewed as a subgroup of the restricted orthogonal group. There are mainly two reasons for doing this. First it is interesting in itself to have a projective representation of the restricted unitary group, then the expression for the Lie algebra cocycle become relatively simple. Second, we will use the theory derived in this section, in Chap. 5 on applications, to construct a representation in the anti-symmetric Fock Hilbert space of the particular loop group LS^1. Furthermore, we will use the restriction of the spin representation to the restricted unitary group to construct a representation of $Diff^+(S^1)$ considered as a subgroup of the (enlarged) restricted unitary group, on a Lie algebra level.

In *section* 2.6 we give the transformation between the particular formalism we use, the formalism of, in Vershiks words, almost linear operators, and the formalism of Araki, among other. The advantage of our formalism is working directly with the physical Hilbert space, the operators, however, being orthogonal operators. The other formalism, advacated by Araki, work with a Hilbert space twice as big as the physical one, but has the advantage of the operators becoming unitarities which commute with a certain complex conjugation operator.

Chapter 3 is devoted to the corresponding symmetric case of the material treated in Chap. 2, following the pattern outlined in Chap. 2. The details in section 3.3 do not appear anywhere else.

In *section* 3.1 we consider the object similar to the CAR-algebra in its Fock representation, however, this object suffers from a lack of norm continuity, hence it is not a C^*-algebra, but only a *-algebra, called the boson algebra. The "exponentialization", by Stones theorem, of the self-adjoint closure of the sum of the creation and annihilation operators in the Fock representation, gives the unitary operators called Weyl operators. These Weyl operators generate the C^*-algebra called the CCR-algebra. The dicussion of this section will be used in the following sections.

In *section* 3.2 the previous section will be used in the construction of second quantization in the symmetric Fock Hilbert space, by methods similar to those of section 2.3. The construction in this section will be used in an essential way in the construction of the metaplectic representation, discussed in the subsequent section.

In *section* 3.3 the infinite dimensional symplectic group is introduced and we construct the metaplectic representation of the restricted metaplectic group, on a Lie algebra level, by use of second quantization and some ideas related to it. It turns out that we may use a strategy similar to that outlined in section 2.4. In particular we state the question: for which symplectic transformations, S, is the automorphism, $\pi(f) \rightarrow \pi(S^{-1}f)$, unitarily implementable in the Fock representation? Similar to the orthogonal case this question has already been answered. Shale's answer from 1962 is that the automorphism is unitarily implementable if and only if S belongs to the restricted symplectic group $Sp_2(\mathcal{H})$. As in the former case our proof is constructive. In fact, these unitary implementers are given by the metaplectic representation of the restricted symplectic group. Beside giving us an explicit projective representation of the restricted symplectic group, this section will be used in the applications of Chap. 5 of, to discuss $Diff^+(S^1)$ realized as a subgroup of the (enlarged) restricted symplectic group, on a Lie algebra level. We end this section by deriving an explicit formula for the vacuum functional in this case.

Chapter 4 is devoted to a general review of the loop algebras and the Virasoro algebra together with certain representations of the Virasoro algebra in physically relevant Hilbert spaces.

In *section* 4.1 we introduce the loop algebra in general. One reason for doing this is to obtain knowledge on loop groups, we will study a particular one in Chap. 5. Central extensions of loop algebras are very important examples of affine Kac-Moody algebras. The simplest representations of affine Kac-Moody algebras are given in terms of the spin representation, whereby the connection between this section and Chap. 2 is clarified. Another reason for studying loop algebras is that the diffeomorphism group $Diff^+(S^1)$, to be studied in the applications in Chap. 5, acts as a group of automorphisms of the central extension of loop algebras, or equivalently at the group level, that the diffeomorphism group $Diff^+(S^1)$ acts as a group of automorphisms of the central extension of loop groups. We end this section by describing how the loop algebras are related to the Virasoro algebra, which is the central extension of the Lie algebra $Vect(S^1)$ of $Diff^+(S^1)$, of smooth vector fields on the unit circle, by the Sugawara construction. It simply gives a representation of the Virasoro algebra by expressing its generators as quadratic terms in the basis elements of any representation of an affine Kac-Moody algebra.

In *section* 4.2 we discuss the Virasoro algebra mentioned above. Its relation to the loop algebras and groups is described in the former section. In Chap. 5 of applications we construct some representations of the Virasoro algebra by use of the spin representation and the

metaplectic representation. However, before doing so, we discuss, in
the following section, some specific representations of the Virasoro
algebra.

In *section* 4.3 we construct some explicit representations of the
Virasoro algebra with central charge $c = \frac{1}{2}$, in the anti-symmetric
Fock Hilbert space. It turns out that there are two essentially different cases: the Ramond sector and the Neveu-Schwarz sector, respectively. These are irreducible unitary lowest weight representations
of positive finite energy. The construction is also an illustration of
how one, in principle, can construct a lot of representations of the
Virasoro algebra for central charge greater than one (and as here,
for $c = \frac{1}{2}$). It is pointed out that the case $c = \frac{1}{2}$ is of special interest,
since it is closely related to the complicated Ising model.

In *section* 4.4 we will construct a series of representations of the
Virasoro algebra with central charge belonging to the discrete series
$c_m = 1 - \frac{6}{(m+2)(m+3)} \in [0,1]$, where $m \in \mathbb{N} \cup \{0\}$. This purely algebraic construction is known as the Goddard-Kent-Olive construc-
tion. As in the former sections we approach the method by more
analytical means, i.e. we expand the known method to constructing representations in a Hilbert space, namely the anti-symmetric
Fock Hilbert space. In particular, we obtain a representation of the
Virasoro algebra with central charge $c = c_m = \frac{1}{2}$, for $m = 1$.

Chapter 5 presents applications of the former chapters and connects the topics considered earlier. The general discussion of the loop
algebras and the Virasoro algebra especially serves as a background
for this chapter. Some of the aspects appearing in this chapter have
been considered briefly by others, we give an elaborated and self-
contained version which, to our knowledge, has not been published
previously.

In *section* 5.1 we use the basic knowledge of loop groups derived
in section 4.1 in discussing the particular loop group LS^1, also called the loop circle. It turns out that it can be decomposed into a
product of the special loop group SLS^1 and the charge group. We
will realize SLS^1 as an abelian subgroup of the restricted unitary
group and then use the spin representation to make explicit representations of SLS^1, on a Lie algebra level. It follows that the spin
representation of the special loop group SLS^1 fulfils the Weyl form
of the canonical commutation relations and thereby becomes a representation of the CCR-algebra, in the anti-symmetric Fock Hilbert
space (indicating the so-called boson-fermion correspondence). Due

to the fact that the elements of the Lie algebra of the charge group fail to be Hilbert-Schmidt when commuted with a certain projection operator, it follows that we cannot use the method developed in section 2.4 or 2.5. So we have to discuss the charge group by other means. We can nevertheless explicitly construct unitary implementers, hence the product of the representation of SLS^1 and the unitary implementers for the charge group, provide us with explicit projective representations of the particular loop group LS^1.

In *section* 5.2 we study the orientation-preserving diffeomorphism group, by realizing it as a subgroup of the restricted unitary group, using the spin representation of the restricted unitary group, on a Lie algebra level. Thereby we get a series of positive energy representations of the Virasoro algebra, with central charge $c = 1$, in terms of the spin representation on the anti-symmetric Fock Hilbert space.

In *section* 5.3 we make considerations analogous to those of section 5.2. However, this time we will consider $Diff^+(S^1)$ as a subgroup of the symplectic group on a particular Hilbert space and thereby we may use the metaplectic representation to obtain a projective positive energy representation, on a Lie algebra level, of the Virasoro algebra with central charge $c = 1$ and lowest weight $h = 0$.

In *section* 5.4 we consider the boson-fermion correspondance, not in general, but in the case discussed in section 5.1. The reason why we have delayed this discussion until now is that we will use some of the considerations arising naturally in the former sections of this chapter.

At the end we have placed a *bibliography*.

Chapter 2

The Spin Representation of the Infinite–Dimensional Orthogonal Group

2.1   The Fock Hilbert Spaces

In this section we describe the construction and interpretation of the Fock Hilbert spaces and some of their important subspaces. It will be widely used in the following.

It was the physicist V. Fock who introduced the Fock Hilbert spaces, as a working ground of quantum field theory, in 1932 ([Fo]), but it was F. Murray and J. von Neumann who first gave a detailed description of finite tensor products of Hilbert spaces, in 1936 ([M-N]), though tensor products of finite dimensional spaces were known long before that. This section is based on [R-S 1], [Fo], [M-N] and [B-R 2].

Let $\mathcal{H}$ denote a separable complex Hilbert space, with inner product $\langle \cdot, \cdot \rangle$, complex linear in the right-hand argument. The vectors in $\mathcal{H}$ describe *the physical one-particle states* of a given quantum physical system. Let $\otimes^n \mathcal{H}$ denote the n-fold (Hilbert space) tensor product of n identical copies of $\mathcal{H}$, for $n \in \mathbf{N}$, where $\otimes^1 \mathcal{H}$ is identified with $\mathcal{H}$. We define $\otimes^0 \mathcal{H} = \mathbf{C}$. Thus, the vectors in $\otimes^n \mathcal{H}$, $n \in \mathbf{N}$, describe *the n-particle quantum physical states*, hence $\otimes^n \mathcal{H}$ is called *the n-particle space*. We now define *the Fock Hilbert space $\mathcal{F}(\mathcal{H})$* as the (Hilbert space) direct sum of the n-fold tensor products $\otimes^n \mathcal{H}$, $n \in \mathbf{N} \cup \{0\}$, i.e.

$$\mathcal{F}(\mathcal{H}) = \oplus_{n=0}^{\infty}(\otimes^n \mathcal{H})$$

providing a canonical grading of $\mathcal{F}(\mathcal{H})$. Thus, a vector $F = \oplus_{n=0}^{\infty} F_n$ of $\mathcal{F}(\mathcal{H})$ can be viewed as a sequence $\{F_n\}_{n=0}^{\infty}$ of vectors $F_n \in \otimes^n \mathcal{H}$ such that $\sum_{n=0}^{\infty} \|F_n\|^2 < \infty$, where the norm is taken in the respective spaces, and

$$\|F_n\| = \|f_1 \otimes \cdots \otimes f_n\| = \|f_1\| \cdot \ldots \cdot \|f_n\|$$

for $F_n \in \otimes^n \mathcal{H}$ of the form $F_n = f_1 \otimes \cdots \otimes f_n$, with $f_1, \ldots, f_n \in \mathcal{H}$. Notice that $\mathcal{F}(\mathcal{H})$ is a Hilbert space with the natural inner product given by $\langle F, G \rangle = \sum_{n=0}^{\infty} \langle F_n, G_n \rangle$, where $F = \{F_n\}_{n=0}^{\infty}$, $G = \{G_n\}_{n=0}^{\infty}$ and $\langle F_n, G_n \rangle = \langle f_1, g_1 \rangle \cdot \ldots \cdot \langle f_n, g_n \rangle$ on product vectors $F_n = f_1 \otimes \cdots \otimes f_n$ and $G_n = g_1 \otimes \ldots \otimes g_n$. The linear span of such product vectors in $\otimes^n \mathcal{H}$, $n \in \mathbf{N} \cup \{0\}$ forms a dense set in $\mathcal{F}(\mathcal{H})$, by definition. We write $F = \oplus_{n=0}^{\infty} F_n$ and $F = \{F_n\}_{n=0}^{\infty}$ interchangeably. The n-particle Hilbert space $\otimes^n \mathcal{H}$ can be identified with the closed subspace of $\mathcal{F}(\mathcal{H})$, consisting of vectors $F = \{F_n\}_{n=0}^{\infty}$ with all components except the n'th equal to zero. Frequently we will consider the closed subspace $\mathcal{D}$ of $\mathcal{F}(\mathcal{H})$ consisting of vectors $F = \oplus_{n=0}^{\infty} F_n$ with only finite many non-zero components F_n, i.e. $\mathcal{D}$ is the algebraic direct sum of the $\otimes^n \mathcal{H}$, and $\mathcal{F}(\mathcal{H})$ is the completion of $\mathcal{D}$ with respect to the Hilbert space norm arising from the inner product. Finally, we define the closed subspace $\mathcal{D}^{(k)}$ of $\mathcal{F}(\mathcal{H})$ consisting of vectors $F = \oplus_{n=0}^{\infty} F_n$ such that F_n is zero for $n > k$, hence we can identify $\mathcal{D}^{(k)}$ with $\oplus_{n=0}^{k} \otimes^n \mathcal{H}$. Notice that $\mathcal{D}^{(k)}$ is also a closed subspace of $\mathcal{D}$. The distinguished vector $\Omega = \oplus_{n=0}^{\infty} \Omega_n$ with $\Omega_0 = 1$ and $\Omega_n = 0$ for all $n \in \mathbf{N}$, is called the *vacuum vector*, since it describes the "empty space" corresponding to no particles.

Actually it is not $\mathcal{F}(\mathcal{H})$ itself, but rather two of its closed subspaces, described below, which are used most frequently. Define two operators on $\otimes^n \mathcal{H}$, *the symmetrization operator* $P_{\vee}$ and *the antisymmetrization operator* $P_{\wedge}$ given on product vectors $f_1 \otimes \cdots \otimes f_n \in \otimes^n \mathcal{H}$ by

$$P_{\divideontimes}(f_1 \otimes \cdots \otimes f_n) = (n!)^{-1} \sum_{\sigma \in \mathcal{S}_n} \chi_{\divideontimes}(\sigma) \cdot f_{\sigma(1)} \otimes \cdots \otimes f_{\sigma(n)}$$

where $P_{\divideontimes}$ denotes either $P_{\vee}$ or $P_{\wedge}$ and the index on $\chi_{\divideontimes}(\sigma)$ corresponds to that of $P_{\divideontimes}$. The sum is taken over all permutations σ in the permutation group $\mathcal{S}_n$ of permutations of n elements, and $\chi_{\wedge}(\sigma)$ is the sign of the permutation σ, whereas $\chi_{\vee}(\sigma)$ is constant 1, and could be omitted. One easily checks, by direct calculations, that both operators $P_{\divideontimes}$ are, in fact, orthogonal projections. We introduce the notation

$$f_1 \wedge \cdots \wedge f_n = (n!)^{\frac{1}{2}} P_{\wedge}(f_1 \otimes \cdots \otimes f_n)$$

$$= (n!)^{-\frac{1}{2}} \sum_{\sigma \in S_n} \chi_\wedge(\sigma) \cdot f_{\sigma(1)} \otimes \cdots \otimes f_{\sigma(n)}$$

and its symmetric analogue

$$f_1 \vee \cdots \vee f_n = (n!)^{\frac{1}{2}} P_\vee(f_1 \otimes \cdots \otimes f_n)$$
$$= (n!)^{-\frac{1}{2}} \sum_{\sigma \in S_n} f_{\sigma(1)} \otimes \cdots \otimes f_{\sigma(n)}$$

Moreover we write $\wedge^n \mathcal{H}$ for $P_\wedge(\otimes^n \mathcal{H})$ and $\vee^n \mathcal{H}$ for $P_\vee(\otimes^n \mathcal{H})$, which clearly are closed subspaces of $\otimes^n \mathcal{H}$. These orthogonal projections are extended in an obvious way to the whole Fock Hilbert space, denoted by the same symbols, by putting

$$P_{\mathbb{X}}(\oplus_{n=0}^\infty F_n) = \oplus_{n=0}^\infty (P_{\mathbb{X}} F_n) \tag{2.1}$$

for $F_n \in \otimes^n \mathcal{H}$. Actually this is done by first extending $P_{\mathbb{X}}$ by linearity, giving two densely defined operators, both with norm 1, then followed by an extension by continuity to bounded operators on $\otimes^n \mathcal{H}$, and finally followed by the extension given by formula (2.1). We write $\mathcal{F}_{\mathbb{X}}(\mathcal{H})$ for $P_{\mathbb{X}} \mathcal{F}(\mathcal{H})$, respectively. Formula (2.1) gives us a natural grading of $\mathcal{F}_{\mathbb{X}}(\mathcal{H})$. Since any product vector of the form $f_1 \vee \cdots \vee f_n$, with $f_1, \ldots, f_n \in \mathcal{H}$, is totally symmetric, i.e. is invariant under all permutations of the vectors $f_1, \ldots, f_n \in \mathcal{H}$, we call $\mathcal{F}_\vee(\mathcal{H})$ *the symmetric Fock Hilbert space* over $\mathcal{H}$ or *the boson Fock Hilbert space*, reflecting the fact that its physical states (vectors) describe bosons (particles obeying Bose-Einstein statistics). Since any product vector of the form $f_1 \wedge \cdots \wedge f_n$, with $f_1, \ldots, f_n \in \mathcal{H}$, is anti-symmetric, i.e. an interchange of any two one-particle vectors in the product vector is equal to minus the original product vector, we call $\mathcal{F}_\wedge(\mathcal{H})$ *the anti-symmetric Fock Hilbert space* over $\mathcal{H}$ or *the fermion Fock Hilbert space*, reflecting the fact that its physical states describe fermions (particles obeying Fermi-Dirac statistics).

Alternatively one could define $\mathcal{F}_\wedge(\mathcal{H})$ as the Hilbert space completion of the exterior algebra over the complex Hilbert space $\mathcal{H}$ and $\mathcal{F}_\vee(\mathcal{H})$ as the Hilbert space completion of the symmetric tensor algebra over the complex Hilbert space $\mathcal{H}$.

The closed subspaces $\mathcal{D}$ and $\mathcal{D}^{(k)}$, discussed earlier, give rise to the corresponding closed subspaces $\mathcal{D}_{\mathbb{X}} = \mathcal{D} \cap \mathcal{F}_{\mathbb{X}}(\mathcal{H})$ and $\mathcal{D}_{\mathbb{X}}^{(k)} = \mathcal{D}^{(k)} \cap \mathcal{F}_{\mathbb{X}}(\mathcal{H})$, in $\mathcal{F}_\vee(\mathcal{H})$ and $\mathcal{F}_\wedge(\mathcal{H})$ respectively. With abuse of notation we suppress the index and simply write $\mathcal{D}$ and $\mathcal{D}^{(k)}$, respectively, when it is clear from the context in which space we are working.

Before ending this section we will discuss the orthonormal basis of the repective spaces.

Let $\{e_k\}_{k=1}^{\infty}$ be an orthonormal basis for $\mathcal{H}$. Then it follows that

$$\{e_{k_1} \otimes \cdots \otimes e_{k_n} : k_1, \ldots, k_n \in \mathbf{N}\}$$

form an orthonormal basis for $\otimes^n \mathcal{H}$, $n \in \mathbf{N}$, with respect to the inner product given earlier on product vectors as

$$\langle f_1 \otimes \cdots \otimes f_n, g_1 \otimes \ldots \otimes g_n \rangle = \prod_{i=1}^{n} \langle f_i, g_i \rangle$$

Hence

$$\{\Omega\} \cup \{e_{k_1} \otimes \cdots \otimes e_{k_n} : k_1, \ldots, k_n \in \mathbf{N}, n \in \mathbf{N}\}$$

is an orthonormal basis for $\mathcal{F}(\mathcal{H})$, where we have identified $e_{k_1} \otimes \cdots \otimes e_{k_n}$ with the vector $E = \oplus_{m=0}^{\infty} E_m \in \mathcal{F}(\mathcal{H})$ such that $E_m = 0$ for all $m \in \mathbf{N} \setminus \{n\}$ and $E_n = e_{k_1} \otimes \cdots \otimes e_{k_n}$.

The restriction of the inner product on $\otimes^n \mathcal{H}$ to $\wedge^n \mathcal{H}$ gives, on product vectors $f_1 \wedge \cdots \wedge f_n$ and $g_1 \wedge \cdots \wedge g_n$, the following

$$
\begin{aligned}
&\langle f_1 \wedge \cdots \wedge f_n, g_1 \wedge \cdots \wedge g_n \rangle \\
&= n! \langle P_\wedge(f_1 \otimes \cdots \otimes f_n), P_\wedge(g_1 \otimes \ldots \otimes g_n) \rangle \\
&= \sum_{\sigma \in \mathcal{S}_n} \chi_\wedge(\sigma) \langle f_{\sigma(1)} \otimes \cdots \otimes f_{\sigma(n)}, g_1 \otimes \ldots \otimes g_n \rangle \\
&= \sum_{\sigma \in \mathcal{S}_n} \chi_\wedge(\sigma) \prod_{i=1}^{n} \langle f_{\sigma(i)}, g_i \rangle \\
&= \det((\langle f_j, g_i \rangle)_{i,j=1,\ldots,n}
\end{aligned}
$$

where $\det((\langle f_j, g_i \rangle)_{i,j=1,\ldots,n}$ denote the determinant of matrix $((\langle f_j, g_i \rangle)_{i,j=1,\ldots,n}$ and we have used that $P_\wedge$ is an orthogonal projection on $\otimes^n \mathcal{H}$. It follows that

$$\{e_{k_1} \wedge \cdots \wedge e_{k_n} : k_1 < \cdots < k_n, k_1, \ldots, k_n \in \mathbf{N}\}$$

form an orthonormal basis for $\wedge^n \mathcal{H}$. Hence, the union of $\{\Omega\}$ and

$$\{e_{k_1} \wedge \cdots \wedge e_{k_n} : k_1 < \cdots < k_n, k_1, \ldots, k_n \in \mathbf{N}, n \in \mathbf{N}\}$$

gives an orthonormal basis for $\mathcal{F}_\wedge(\mathcal{H})$, where we have used the canonical identification of vectors in $\wedge^n \mathcal{H}$ with vectors in $\mathcal{F}_\wedge(\mathcal{H})$, in analogy with the above.

The restriction of the inner product to $\mathcal{F}_V(\mathcal{H})$ gives

$$
\begin{aligned}
\langle f_1 &\vee \cdots \vee f_n, g_1 \vee \cdots \vee g_n \rangle \\
&= n! \langle P_V(f_1 \otimes \cdots \otimes f_n), P_V(g_1 \otimes \ldots \otimes g_n) \rangle \\
&= \sum_{\sigma \in \mathcal{S}_n} \langle f_{\sigma(1)} \otimes \cdots \otimes f_{\sigma(n)}, g_1 \otimes \ldots \otimes g_n \rangle \\
&= \sum_{\sigma \in \mathcal{S}_n} \prod_{i=1}^{n} \langle f_{\sigma(i)}, g_i \rangle
\end{aligned}
$$

known as the permanent of matrix $(\langle f_j, g_i \rangle)_{i,j=1,\ldots,n}$, where we have used that P_V is an orthogonal projection on $\otimes^n \mathcal{H}$. If $(e_{k_1}, \ldots, e_{k_n}) = (e_{\sigma(l_1)}, \ldots, e_{\sigma(l_n)})$ for exactly m different permutations $\sigma \in \mathcal{S}_n$, where $m \in \mathbf{N} \cup \{0\}$, then

$$
\langle e_{k_1} \vee \cdots \vee e_{k_n}, e_{l_1} \vee \cdots \vee e_{l_n} \rangle = \sum_{\sigma \in \mathcal{S}_n} \prod_{i=1}^{n} \langle e_{k_i}, e_{\sigma(l_i)} \rangle = m
$$

If we collect the factors and write $e_{k_i}^{r_i}$ for $\overbrace{e_{k_i} \vee \cdots \vee e_{k_i}}^{r_i\text{-times}}$, then

$$
\langle e_{k_1}^{r_1} \vee \cdots \vee e_{k_n}^{r_n}, e_{k_1}^{r_1} \vee \cdots \vee e_{k_n}^{r_n} \rangle = \prod_{i=1}^{n} (r_i!)
$$

and

$$
\langle e_{k_1}^{r_1} \vee \cdots \vee e_{k_p}^{r_p}, e_{l_1}^{s_1} \vee \cdots \vee e_{l_m}^{s_m} \rangle = 0
$$

if $e_{k_1}^{r_1} \vee \cdots \vee e_{k_p}^{r_p} \neq e_{l_1}^{s_1} \vee \cdots \vee e_{l_m}^{s_m}$, where $n = r_1 + \cdots + r_p = s_1 + \cdots + s_m$ and the inner product is taken in $\vee^n \mathcal{H}$. Then

$$
\left\{ \prod_{i=1}^{n} (r_i!)^{-\frac{1}{2}} \cdot e_{k_1}^{r_1} \vee \cdots \vee e_{k_n}^{r_n} \quad : \quad k_1 < \cdots < k_n, \right.
$$
$$
\left. k_1, \ldots, k_n, r_1, \ldots, r_n \in \mathbf{N} \right\}
$$

form an orthonormal basis for $\vee^n \mathcal{H}$. Hence

$$
\{\Omega\} \cup \left\{ K_{(r)_n} \cdot e_{k_1}^{r_1} \vee \cdots \vee e_{k_n}^{r_n} \quad : \quad k_1 < \cdots < k_n, \right.
$$
$$
\left. (k)_n, (r)_n \in \mathbf{N}^n, n \in \mathbf{N} \right\}
$$

is an orthonormal basis for $\mathcal{F}_V(\mathcal{H})$, where we have used multiindex notation, $(k)_n = (k_1, \ldots, k_n)$ and $(r)_n = (r_1, \ldots, r_n)$ and the canonical identification of vectors in $\vee^n \mathcal{H}$ with the corresponding vectors in $\mathcal{F}_\vee(\mathcal{H})$, in analogy with earlier. The normalization constant is given by

$$K_{(r)_n} = \|e_{k_1}^{r_1} \vee \cdots \vee e_{k_n}^{r_n}\|^{-1} = \prod_{i=1}^n (r_i!)^{-\frac{1}{2}}$$

We will sometimes use the notation of multiindex, as mentioned above, writing $(k)_n$ for $(k_1, \ldots, k_n)$ for $n \in \mathbf{N}$, where $k_i \in \mathbf{N}$ or sometimes $k_i \in \mathbf{N} \cup \{0\}$ for $i = 1, \ldots, n$.

With the detailed knowledge of the structure of the Fock Hilbert spaces we are able to discuss the so-called second quatization and related topics in detail, but first we turn to the discussion of the CAR-algebra and its Fock representation, where CAR-algebra stands for canonical anti-commutation relations.

2.2   The CAR-Algebra and its Fock Representation

In this section we describe the CAR-algebra, which is a C^*-algebra, unique up to $*$-isomorphism. The CAR-algebra was introduced by Jordan and Wigner ([J-W]) in 1928 for the purpose of quantization of the electron field in physics. There is a very useful concrete realization of the CAR-algebra on the anti-symmetric Fock Hilbert space, called the Fock representation, which is the one physicists generally use. It was this representation of the CAR-algebra that J.M. Cook, a student of I. E. Segal, used in his fundamental, and remarkably detailed, paper ([Co]) in 1953. However, the basic idea goes back to V. Fock ([Fo]), in 1932, and Jordan and Wigner ([J-W]), in 1928.

This section gives us a good mathematical frame for handling many problems in quantum physics with infinitely many degrees of freedom. The topic of this section is also treated in [Co], [Ar] and [B-R 2].

Let $\mathcal{H}$ be a Hilbert space over $\mathbf{C}$, and $\mathcal{A}$ an abstract C^*-algebra with a unit. We call $\mathcal{A}$ a CAR-*algebra over* $\mathcal{H}$ if there is an anti-linear mapping $a : \mathcal{H} \to \mathcal{A}$ such that $\{a(f) : f \in \mathcal{H}\}$, together with the unit I, generate the C^*-algebra $\mathcal{A}$ and fulfil the *canonical anti-commutation relations*

$$\begin{aligned}[a(f), a(g)]_+ &= 0 \\ [a(f), a(g)^*]_+ &= \langle f, g \rangle \cdot I\end{aligned} \qquad (2.2)$$

for all $f, g \in \mathcal{H}$. Here $\langle \cdot, \cdot \rangle$ denotes the inner product on $\mathcal{H}$ and $[\cdot, \cdot]_+$ the usual anti-commutator $[A, B]_+ = AB + BA$. We will refer to the canonical anti-commutation relations, given by (2.2), as the CAR.

By use of the CAR, it follows that $f \to a(f)$ is an isometry, since

$$(a(f)^*a(f))^2 = a(f)^* \left(\|f\|^2 \cdot I - a(f)^*a(f) \right) a(f) = \|f\|^2 \cdot a(f)^*a(f)$$

so

$$\|a(f)\|^4 = \|(a(f)^*a(f))^2\| = \|f\|^2 \cdot \|a(f)^*a(f)\| = \|f\|^2 \cdot \|a(f)\|^2$$

hence $\|a(f)\| = \|f\|$, where we have used the same norm-symbol to denote different norms, one acting on the C^*-algebra and the other on the Hilbert space. Moreover, a CAR-algebra is unique up to $*$-isomorphism. To prove this, let $\tilde{A}$ be another CAR-algebra over the same Hilbert space $\mathcal{H}$ and let $\tilde{a} : \mathcal{H} \to \tilde{A}$ denote the corresponding anti-linear mapping. Define a $*$-morhpism (i.e. a linear, multiplicative mapping conserving the involution)

$$\alpha : \{a(f), a(f)^* : f \in \mathcal{H}\} \to \{\tilde{a}(f), \tilde{a}(f)^* : f \in \mathcal{H}\}$$

by

$$\alpha(a(f)) = \tilde{a}(f)$$

for all $f \in \mathcal{H}$. Clearly α is an isometry on its domain, hence it can be extended, in a well-defined manner, to the C^*-algebra $\mathcal{A}$ generated by $\{a(f) : f \in \mathcal{H}\}$. Moreover its range is the C^*-subalgebra of $\tilde{A}$, generated by $\{\tilde{a}(f), \tilde{a}(f)^* : f \in \mathcal{H}\}$, i.e. the range is all of $\tilde{A}$. Then the $*$-morphism α is one-to-one and onto, hence a $*$-isomorphism. So we may speak of *the CAR-algebra*, since it is unique up to $*$-isomorphism.

Let U be any bounded complex linear operator on $\mathcal{H}$, and V any bounded complex anti-linear operator satisfying

$$V^*U + U^*V = 0 = UV^* + VU^*$$

and

$$U^*U + V^*V = I = UU^* + VV^* \tag{2.3}$$

we then say that U and V are *Bogoliubov transformations*. Notice that the adjoint of an anti-linear operator V is defined in a different way to the adjoint of a linear operator. We have $\langle f, Vg \rangle = \langle g, V^*f \rangle = \overline{\langle V^*f, g \rangle}$ for all $f, g \in \mathcal{H}$; for a more detailed discussion we refer to section 2.4. Put $\tilde{a}(f) = a(Uf) + a(Vf)^*$. Then $\tilde{a}(f)$ is evidently bounded for all $f \in \mathcal{H}$. It follows that the mapping $\tilde{a} : f \to \tilde{a}(f)$ is anti-linear, which implies that

$$
\begin{aligned}
[\tilde{a}(f), \tilde{a}(g)]_+ &= \langle Uf, Vg \rangle \cdot I + \overline{\langle Vf, Ug \rangle} \cdot I \\
&= \langle g, (V^*U + U^*V)f \rangle \cdot I \\
&= 0
\end{aligned}
$$

and that

$$
\begin{aligned}
[\tilde{a}(f), \tilde{a}(g)^*]_+ &= \langle Uf, Ug \rangle \cdot I + \langle Vf, Vg \rangle \cdot I \\
&= \langle (U^*U + V^*V)f, g \rangle \cdot I \\
&= \langle f, g \rangle \cdot I
\end{aligned}
$$

for all $f, g \in \mathcal{H}$, where we have used the definition of the adjoint of an anti-linear operator V. So $\{\tilde{a}(f) : f \in \mathcal{H}\} \cup \{I\}$ generates a CAR-algebra over $\mathcal{H}$. Hence, because of the uniqueness, up to $*$-isomorphism, of the CAR-algebra, there exists a unique $*$-automorphism α of the CAR-algebra, such that $\alpha(a(f)) = \tilde{a}(f)$, for all $f \in \mathcal{H}$, and in this case is $\alpha^{-1}(\tilde{a}(f)) = a(U^*f) + a(V^*f)^*$, which follows directly by calculation.

It is easily proved that the CAR-algebra is separable if and only if $\mathcal{H}$ is separable and that the CAR-algebra is simple (for proofs, see [B-R 2, p. 16]). In our case we will only consider infinite dimensional (complex or real) separable Hilbert spaces.

We now turn to *the Fock representation of the CAR-algebra*. For each $f \in \mathcal{H}$ define the linear operator $a_0{}^*(f)$ on the anti-symmetric Fock Hilbert space $\mathcal{F}_\wedge(\mathcal{H})$ by

$$
a_0{}^*(f)\Omega = f
$$

and on product vectors by

$$
a_0{}^*(f)(f_1 \wedge \cdots \wedge f_n) = f \wedge f_1 \wedge \cdots \wedge f_n
$$

Notice that $a_0{}^*(f) : \wedge^n \mathcal{H} \to \wedge^{n+1} \mathcal{H}$. Extension by linearity yields a densely defined linear operator on $\mathcal{F}_\wedge(\mathcal{H})$. For $f \neq 0$ choose an orthonormal basis $\{e_i\}_{i \in \mathbf{N}}$ for $\mathcal{H}$ such that $e_1 = \frac{f}{\|f\|}$. One then observes that $a_0{}^*(e_1) = \|f\|^{-1} a_0{}^*(f)$ defines a partial isometry with support

$$
\mathcal{K} = \overline{span}\{\Omega, e_{i_1} \wedge \cdots \wedge e_{i_n} : 1 < i_1 < \cdots < i_n, i_1, \ldots, i_n, n \in \mathbf{N}\}
$$

and range $\mathcal{K}^{\perp} \cap \mathcal{F}_{\wedge}(\mathcal{H})$, since $a_0^*(e_1)(e_1 \wedge e_{i_2} \wedge \cdots \wedge e_{i_n}) = 0$ for all $i_2 < \cdots < i_n$ and $\|a_0^*(e_1)(e_{i_1} \wedge \cdots \wedge e_{i_n})\| = \|e_1 \wedge e_{i_1} \wedge \cdots \wedge e_{i_n}\| = 1$ for all $1 < i_1 < \cdots < i_n$. Hence $\|a_0^*(f)\| = \|f\|$ for all $f \in \mathcal{H}$. So $a_0^*(f)$ is a bounded operator on $\mathcal{F}_{\wedge}(\mathcal{H})$, hence it has a unique bounded adjoint $a_0^*(f)^*$, which we denote $a_0(f)$. It is given by

$$a_0(f)\Omega = 0$$

and

$$
\begin{aligned}
&a_0(f)(f_1 \wedge \cdots \wedge f_n) \\
&= \sum_{i=1}^{n}(-1)^{i+1}\langle f, f_i\rangle \cdot f_1 \wedge \cdots \wedge f_{i-1} \wedge f_{i+1} \wedge \cdots \wedge f_n
\end{aligned}
$$

since

$$
\begin{aligned}
&\langle a_0^*(g_1)(g_2 \wedge \cdots \wedge g_n), f_1 \wedge \cdots \wedge f_n\rangle \\
&= \langle g_1 \wedge g_2 \wedge \cdots \wedge g_n, f_1 \wedge \cdots \wedge f_n\rangle \\
&= \det\{\langle g_i, f_j\rangle\}_{i,j=1,\ldots,n}
\end{aligned}
$$

and expansion after the first row gives

$$
\sum_{i=1}^{n}(-1)^{i+1}\langle g_1, f_i\rangle \cdot \langle g_2 \wedge \cdots \wedge g_n, f_1 \wedge \cdots \wedge f_{i-1} \wedge f_{i+1} \wedge \cdots \wedge f_n\rangle
$$

$$
= \langle g_2 \wedge \cdots \wedge g_n, \sum_{i=1}^{n}(-1)^{i+1}\langle g_1, f_i\rangle \cdot f_1 \wedge \cdots \wedge f_{i-1} \wedge f_{i+1} \wedge \cdots \wedge f_n\rangle
$$

which gives the above formula for $a_0(f) = a_0^*(f)^*$ on product vectors, and evidently is $0 = \langle a_0^*(f)F_n, \Omega\rangle = \langle F_n, a_0(f)\Omega\rangle$, for any $F_n \in \wedge^n \mathcal{H}$ and for all $n \in \mathbb{N} \cup \{0\}$, so $a_0(f)\Omega = 0$. Observe that $\|a_0(f)\| = \|a_0^*(f)\| = \|f\|$. Moreover, notice that $a_0^*(f) : \wedge^n \mathcal{H} \to \wedge^{n+1} \mathcal{H}$ and that $a_0(f) : \wedge^{n+1} \mathcal{H} \to \wedge^n \mathcal{H}$, hence their names, *creation* and *annihilation operators*, respectively, since they correspond to creation and annihilation of particles in quantum physics. Observe now that

$$a_0(f)a_0^*(f) + a_0^*(f)a_0(f) = \|f\|^2 \cdot I$$

first on Ω and on product vectors, then by linearity and continuity on all of $\mathcal{F}_{\wedge}(\mathcal{H})$. From this it follows, again, that $a_0^*(f)$ and $a_0(f)$ are bounded by $\|f\|$, since both terms on the left-hand side are positive. Moreover

$$a_0{}^*(f)a_0{}^*(f) + a_0{}^*(f)a_0{}^*(f) = 0$$

for all $f \in \mathcal{H}$, due to the anti-symmetry of product vectors in $\mathcal{F}_\wedge(\mathcal{H})$. The last equation reflects the famous Pauli principle from physics. Taking the adjoint, we obtain

$$a_0(f)a_0(f) + a_0(f)a_0(f) = 0$$

for all $f \in \mathcal{H}$. Using the polarization identity we get the CAR

$$[a_0(f), a_0{}^*(g)]_+ = \langle f, g \rangle \cdot I$$

and

$$[a_0(f), a_0(g)]_+ = 0 = [a_0{}^*(f), a_0{}^*(g)]_+$$

Hence the concrete C^*-algebra generated by $\{a_0(f) : f \in \mathcal{H}\} \cup \{I\}$ is a representation of the CAR-algebra, called *the Fock representation of the CAR-algebra*, and sometimes physicists denote it *the fermionic field algebra over* $\mathcal{H}$, for obvious reasons. Notice that the vacuum vector Ω is a cyclic vector for the Fock representation of the CAR-algebra.

Now, let P be an arbitrary orthogonal projection in $\mathcal{H}$. Choose an orthonormal basis $\{e_i\}_{i \in \mathbf{N}}$ for $\mathcal{H}$ consisting of eigenvectors for P. Then any $f \in \mathcal{H}$ has an expansion, $f = \sum_{i \in \mathbf{N}} \langle e_i, f \rangle e_i$. We may define an operator Γ on $\mathcal{H}$ given by $\Gamma f = \sum_{i \in \mathbf{N}} \overline{\langle e_i, f \rangle} e_i$, where the bar denotes ordinary complex conjugation. Hence Γ is a well-defined involution on $\mathcal{H}$ and evidently commutes with P. Moreover, Γ is anti-unitary, $\langle \Gamma f, \Gamma g \rangle = \langle g, f \rangle$, for all $f, g \in \mathcal{H}$. That is, for any orthogonal projection P on $\mathcal{H}$ there exists at least one anti-unitary involution Γ on $\mathcal{H}$ such that $[P, \Gamma] = 0$. Let Γ be any such anti-unitary involution on $\mathcal{H}$. Putting $U = I - P$ and $V = \Gamma P$ we see that U and V are Bogoliubov transformations

$$U^*U + V^*V = I - P + P = I = UU^* + VV^*$$

and

$$V^*U + U^*V = \Gamma P(I - P) + (I - P)\Gamma P = 0 = UV^* + VU^*$$

and both U and V are selfadjoint. Then it follows that the C^*-algebra generated by

$$a_P(f) = a_0((I - P)f) + a_0{}^*(\Gamma P f)$$

defines a representation of the CAR-algebra in the Fock space, called the quasi-free representation, in terms of the Fock representation. Observe that this represents the physical idea of "filling up the Dirac sea", since $a_P(f_+)\Omega = 0$, for $f_+ \in (I - P)\mathcal{H}$ (reflecting the fact that the physical states with positive energy are all unoccupied) and $a_P{}^*(f_-)\Omega = 0$, for $f_- \in P\mathcal{H}$ (reflecting the fact that the physical states with negative energy are all occupied). So Ω corresponds to the filled Dirac sea in the case of the quasi-free representation. Note that the definition is consistent for $P = 0$, so the Fock representation is just a special case of this construction. Moreover, Ω is a cyclic vector for this representation of the CAR-algebra. Whence we have the following theorem.

Theorem 1 *The representations of the CAR-algebra given by $a(f) \rightarrow a_P(f)$ are all irreducible, including the Fock representation.*

Proof. In the case of $P = 0$ the proof appears in [B-R 2, proposition 5.2.2]. We will repeat and use it below. Since $a_0(f) = a_P((I - P)f) + a_P^*(-\Gamma P f)$ and $a_0^*(f) = a_P^*((I - P)f) + a_P(-\Gamma P f)$, any operator T commuting with $a_P(f)$ and $a_P^*(f)$, $f \in \mathcal{H}$, commutes with $a_0(f)$ and $a_0^*(f)$, and conversely. Therefore we only need to consider the statement for $P = 0$, since it implies the case when $P \neq 0$. Let T commute with $a_0(f)$ and $a_0^*(f)$, i.e. $[T, a_0(f)] = [T, a_0^*(f)] = 0$, Observe that

$$
\begin{aligned}
T_{n,m} &= \langle a_0^*(f_1) \cdot \ldots \cdot a_0^*(f_n)\Omega, T a_0^*(g_1) \cdot \ldots \cdot a_0^*(g_m)\Omega \rangle \\
&= \langle T^*\Omega, a_0(f_n) \cdot \ldots \cdot a_0(f_1) a_0^*(g_1) \cdot \ldots \cdot a_0^*(g_m)\Omega \rangle
\end{aligned}
$$

is zero for $n > m$ and $n, m \in \mathbf{N}$, since we annihilate more particles than we create from Ω. Analogously, we get

$$
\begin{aligned}
T_{n,m} &= \langle a_0(g_m) \cdot \ldots \cdot a_0(g_1) a_0^*(f_1) \cdot \ldots \cdot a_0^*(f_n)\Omega, T\Omega \rangle \\
&= 0
\end{aligned}
$$

for $n < m$. Moreover, for $n = m$, we have $a_0(f_n) \cdot \ldots \cdot a_0(f_1) a_0^*(g_1) \cdot \ldots \cdot a_0^*(g_m)\Omega = b \cdot \Omega$, where

$$
\begin{aligned}
b &= \langle \Omega, a_0(f_n) \cdot \ldots \cdot a_0(f_1) a_0^*(g_1) \cdot \ldots \cdot a_0^*(g_m)\Omega \rangle \\
&= \langle a^*(f_1) \cdot \ldots \cdot a^*(f_n)\Omega, a^*(g_1) \cdot \ldots \cdot a^*(g_m)\Omega \rangle
\end{aligned}
$$

for any $n = m \in \mathbf{N}$. So with $c = \langle T^*\Omega, \Omega \rangle = \langle \Omega, T\Omega \rangle$ we get

$$
T_{n,m} = c \cdot \langle a^*(f_1) \cdot \ldots \cdot a^*(f_n)\Omega, a^*(g_1) \cdot \ldots \cdot a^*(g_m)\Omega \rangle
$$

in all cases, $n, m \in \mathbf{N}$. Trivially, $T_{n,0} = T_{0,n} = \delta_n \cdot c$, so for any $n, m \in \mathbf{N} \cup \{0\}$ we have

$$\langle f_1 \wedge \cdots \wedge f_n, (T - c \cdot I) g_1 \wedge \cdots \wedge g_m \rangle = 0$$

where $f_1 \wedge \cdots \wedge f_n$ and $g_1 \wedge \cdots \wedge g_m$ means Ω for n and m zero, respectively. Hence, $T = c \cdot I$ on all of $\mathcal{H}$, since product vectors span $\mathcal{F}_\wedge(\mathcal{H})$. Then due to a well-known theorem (see for example [B-R 1, p.47]) it follows that the representation is irreducible, proving the claim.

$\square$

The CAR-algebra has an equivalent formulation, in terms of the *Clifford algebra*, which is the analogue description to that of the CCR-algebra (see Chap. 3).

Define $\pi(f)$ by $\pi(f) = \frac{1}{\sqrt{2}} (a(f) + a(f)^*)$. Then $a(f)$ and $a(f)^*$ can be recovered from the $\pi(f)$ by the formula $a(f) = \frac{1}{\sqrt{2}} (\pi(f) + i\pi(if))$ and $a(f)^* = \frac{1}{\sqrt{2}} (\pi(f) - i\pi(if))$. Moreover, it follows from the CAR that

$$
\begin{aligned}
[\pi(f), \pi(g)]_+ &= \frac{1}{2}([a(f), a(g)]_+ + [a(f), a(g)^*]_+ \\
&\quad + [a(f)^*, a(g)]_+ + [a(f)^*, a(g)^*]_+) \\
&= \frac{1}{2}(\langle f, g \rangle + \overline{\langle f, g \rangle}) \cdot I = \operatorname{Re}\langle f, g \rangle \cdot I \\
&= \tau(f, g) \cdot I
\end{aligned}
$$

where $\tau(f, g) = \operatorname{Re}\langle f, g \rangle$ is a real positive symmetric bilinear form on $\mathcal{H}$. Conversely, the CAR can be recovered from the relations $[\pi(f), \pi(g)]_+ = \tau(f, g) \cdot I$. *The orthogonal group* $\mathcal{O}(\mathcal{H})$ consists of those real linear invertible mappings $T : \mathcal{H} \to \mathcal{H}$ such that $\tau(Tf, Tg) = \tau(f, g)$, for all $f, g \in \mathcal{H}$. Consider now the complex Clifford algebra over $\mathcal{H}$, as a real Hilbert space. Then the above relations become $a(f) = \frac{1}{\sqrt{2}} (\pi(f) + i\pi(Jf))$ and $a(f)^* = \frac{1}{\sqrt{2}} (\pi(f) - i\pi(Jf))$, where we have introduced a complex structure on the real Hilbert space through the operator J (this is done in detail in the beginning of section 2.4). For each $T \in \mathcal{O}(\mathcal{H})$, define $\pi_T(f)$ by $\pi_T(f) = \pi(T^{-1}f)$, then $[\pi_T(f), \pi_T(g)]_+ = \tau(f, g)I$. Thus, the mapping $\pi(f) \to \pi_T(f)$ defines an automorphism of the Clifford algebra, and these automorphisms form an automorphism group. It follows (see section 1.4 below) that these orthogonal transformations can be split into a sum $T = U + V$, where U and V are Bogoliubov transformations (see formula (2.3) and compare with formulæ (2.5) and (2.6) in section 2.4) corresponding to the above mentioned equivalence of the CAR-algebra and the Clifford algebra.

At this point it is natural to ask the question: For which $T \in \mathcal{O}(\mathcal{H})$ is this automorphism unitarily implementable in the Fock representation, i.e. for which $T \in \mathcal{O}(\mathcal{H})$ does there exists a unitary operator U_T on $\mathcal{F}_\wedge(\mathcal{H})$ such that $\pi_T(f) = U_T^{-1}\pi(f)U_T$ for all $f \in \mathcal{H}$? This question has been answered by Shale and Stinespring in [S-S] and is covered in detail by Araki in [Ar] and Lundberg in [Lu 2]. We return to this question in section 2.4. But first we will discuss the second quantization, and some of its generalizations.

2.3   The Second Quantization in $\mathcal{F}_\wedge(\mathcal{H})$

In this section we will describe the second quantization, based on the explicit Fock Hilbert space construction in the anti-symmetric case, given in section 2.1, and on the Fock representation of the CAR-algebra, given in section 2.2. This section will be used frequently in the construction of a special representation of the *spin* algebra and is therefore essential for the rest of this book.

The basic idea of this section goes back to V. Fock ([Fo]), in 1932. But it was J.M. Cook, supervised by I. E. Segal, who constructed the method called second quantization in detail ([Co]), in 1953. The method gave a neat mathematical frame for handling many problems in quantum physics with infinitely many degrees of freedom. For another excellent treatment of second quantization we refer to the book of F. A. Berezin, [Be], from 1966. The method of second quantization is, briefly stated, the method of lifting one-particle operators, on a Hilbert space, to many-particle operators, on the Fock Hilbert spaces, whenever it is possible (in the case of the CAR-algebra, one use the anti-symmetric Fock Hilbert space). The method of second quantization has been generalized by Araki in [Ar], Lundberg in [Lu 1] and others, in the sense that they lift trace-class operators and Hilbert-Schmidt operators, fulfilling some additional conditions, to the CAR-algebra or to operators acting in the Fock Hilbert spaces. Our method of constructing the second quantization is different from that given by Cook in [Co].

Let A be a skew-self-adjoint linear operator in $\mathcal{H}$, i.e. $A^* = -A$. Suppose for a while that A is a bounded operator, we may define the operators $U(A)_n$ acting on Ω and on product vectors in $\wedge^n \mathcal{H}$ (for arbitrary $n \in \mathbf{N}$) by $U(A)_0 \Omega = \Omega$ and

$$U(A)_n(f_1 \wedge \cdots \wedge f_n) = e^A f_1 \wedge \cdots \wedge e^A f_n$$

Notice that

$$
\begin{aligned}
& \langle U(A)_n(f_1 \wedge \cdots \wedge f_n), U(A)_n(g_1 \wedge \cdots \wedge g_n) \rangle \\
&= \det\{\langle e^A f_i, e^A g_j \rangle\}_{i,j=1,\ldots,n} \\
&= \det\{\langle f_i, g_j \rangle\}_{i,j=1,\ldots,n} \\
&= \langle f_1 \wedge \cdots \wedge f_n, g_1 \wedge \cdots \wedge g_n \rangle
\end{aligned}
$$

Extension by linearity and continuity gives that each $U(A)_n$ is a well-defined unitary operator on $\wedge^n \mathcal{H}$. Then $U(tA)_n$ is a strongly continuous one-parameter unitary group (see for example [R-S 1, p. 265]) of operators on $\wedge^n \mathcal{H}$, $t \in \mathbf{R}$. Hence, by a transformation of Stone's theorem (see [R-S 1, p. 266]) to skew-self-adjoint operators, there exists a skew-self-adjoint operator $dU(A)_n$ on $\wedge^n \mathcal{H}$ such that

$$U(tA)_n = e^{t \cdot dU(A)n}$$

$t \in \mathbf{R}$. The closed densely defined operator $dU(A)_n$ is called the infinitesimal generator, or just the generator, of $U(A)_n$, since

$$dU(A)_n = \left. \frac{d}{dt} \right|_{t=0} U(tA)_n$$

on its domain, consisting of those vectors $F_n \in \wedge^n \mathcal{H}$ for which the limit of $t^{-1}(U(tA)_n - I)F_n$ exists, as $t \to 0$. Let $F_n = f_1 \wedge \cdots \wedge f_n$ be an arbitrary product vector in $\wedge^n \mathcal{H}$, then

$$dU(A)_n F_n = \sum_{i=1}^{n} f_1 \wedge \cdots \wedge f_{i-1} \wedge A f_i \wedge f_{i+1} \wedge \cdots \wedge f_n \qquad (2.4)$$

which is a well-defined finite linear combination of n-particle product vectors in $\wedge^n \mathcal{H}$. Observe now, that the algebraic direct sum $\mathcal{D} = \oplus_{alg} \wedge^n \mathcal{H}$ equals $\mathcal{D}_0 = \mathcal{A}_0 \Omega$, where $\mathcal{A}_0$ is the C^*-algebra generated by $\{a_0(f) : f \in \mathcal{H}\}$, which is the Fock space realization of the CAR-algebra. Since $U(tA)_0 \Omega = \Omega$ is independent of t, it follows that $dU(A)_0 \Omega = 0$. We may extend $dU(A)_n$ to $\mathcal{D}$. Hence, $dU(A) = \oplus_{n=0}^{\infty} dU(A)_n$ is a well-defined skew-symmetric operator, with dense invariant domain $\mathcal{D}$. Now, consider a possible unbounded skew-self-adjoint operator A. We then define $dU(A)_n$ directly by (2.4) on product vectors in $\wedge^n \mathcal{H}$, such that each one-particle vector, in the product vector, belongs to $\mathcal{D}(A)$, we denote these vectors by $\mathcal{D}(A)^{\wedge n}$, for each $n \in \mathbf{N}$, and $dU(A)_0 \Omega = 0$. Put $dU(A) = \oplus_{n=0}^\infty dU(A)_n$ on $\oplus_{alg} \mathcal{D}(A)^{\wedge n}$.

We will now show that $dU(A)$ has a dense set of analytic vectors (both with regard to the norm topology and the week topology), in the case of a general skew-self-adjoint operator A. First, a direct calculation gives

$$dU(A)(f_1 \wedge \cdots \wedge f_n) = (n!)^{\frac{1}{2}} P_\wedge dU(A)_\otimes (f_1 \otimes \cdots \otimes f_n)$$

for $f_1, \ldots, f_n \in \mathcal{D}(A)$, where

$$dU(A)_\otimes (f_1 \otimes \cdots \otimes f_n) = \sum_{i=1}^n f_1 \otimes \cdots \otimes f_{i-1} \otimes A f_i \otimes f_{i+1} \otimes \cdots \otimes f_n$$

and in the same manner we get

$$dU(A)^k (f_1 \wedge \cdots \wedge f_n) = (n!)^{\frac{1}{2}} P_\wedge dU(A)_\otimes^k (f_1 \otimes \cdots \otimes f_n)$$

Now if each f_j, in the product vector, is an analytic vector for A, then there exists a $M_j < \infty$, for each f_j, such that $\|Af_j\| \le M_j \|f_j\| < \infty$. The skew-self-ajointness of A implies that $\mathcal{D}(A)$ has a dense set of analytic vectors. For product vectors of analytic vectors for A we have

$$\|dU(A)_\otimes (f_1 \otimes \cdots \otimes f_n)\|^2$$
$$= \sum_{\substack{i,j=1 \\ i \neq j}}^n \langle Af_i, f_i \rangle \langle f_j, Af_j \rangle \prod_{\substack{m=1 \\ m \neq i,j}}^n \langle f_m, f_m \rangle + \sum_{i=1}^n \langle Af_i, Af_i \rangle \prod_{\substack{m=1 \\ m \neq i}}^n \langle f_m, f_m \rangle$$
$$\le M^2 \sum_{i,j=1}^n \prod_{m=1}^n \|f_m\|^2 = n^2 \cdot M^2 \cdot \|f_1 \otimes \cdots \otimes f_n\|^2$$

where $M = \max\{M_1, \ldots, M_n\}$. Hence

$$\|dU(A)^k (f_1 \wedge \cdots \wedge f_n)\| \le (n!)^{\frac{1}{2}} \|dU(A)_\otimes^k (f_1 \otimes \cdots \otimes f_n)\|$$
$$\le (n!)^{\frac{1}{2}} n^k M^k \|f_1 \otimes \cdots \otimes f_n\|$$

and then, for any n-particle product vector $f_1 \wedge \cdots \wedge f_n$ such that each f_j is an analytic vector for A

$$\sum_{k=0}^{\infty} \frac{1}{k!} \|dU(A)^k(f_1 \wedge \cdots \wedge f_n)\| \leq \sum_{k=0}^{\infty} \frac{(n \cdot M)^k}{k!} \|f_1 \otimes \cdots \otimes f_n\|(n!)^{\frac{1}{2}}$$

$$= (n!)^{\frac{1}{2}} e^{n \cdot M} \|f_1 \otimes \cdots \otimes f_n\|$$

$$< \infty$$

Which means that the set of finite linear combination of Ω and n-particle product vectors $f_1 \wedge \cdots \wedge f_n$, $n \in \mathbf{N}$, such that each vector f_j is an analytic vector for A, forms a dense set of analytic vectors for $dU(A)$. We denote the set of all analytic vectors for $dU(A)$ by $\mathcal{D}_A$ (for an alternative proof, see [R-S 2, p. 205]). Notice that in case of A bounded, the above M and all M_j may be chosen equal to $\|A\|$, so $\mathcal{D}_A$ becomes all of $\mathcal{D}$ (any vector for a bounded opera-tor A is trivially an analytic vector for A). Then a transformation of Nelson's theorem (see for example [R-S 2, p. 202]) to essentially skew-self-adjoint operators, states that the operator $dU(A)$ is essen-tially skew-self-adjoint (which can also be seen by a quite different argument given, for example, in [R-S 1, p. 302]). Hence, the closure of $dU(A)$, which we also denote by $dU(A)$, is skew-self-adjoint, and generates a strongly continuous one-parameter unitary group, by a transformation of Stone's theorem to skew-self-adjoint operators. We denote this strongly continuous unitary one-parameter group by $U(tA)$.

The above mapping $A \rightarrow dU(A)$, mapping skew-self-adjoint ope-rators on $\mathcal{H}$ into skew-self-adjoint operators on $\mathcal{F}_\wedge(\mathcal{H})$, is called *the second quantization* mapping.

Theorem 2 *The mapping of second quantization, $A \rightarrow dU(A)$, on skew-self-adjoint operators A in $\mathcal{H}$ fulfils*

1) $U(tA)a_0(f)U(-tA) = a_0(e^{tA}f)$, *for analytic vectors f for A and $t \in \mathbf{R}$.*

2) $\overline{[dU(A), a_0^*(f)]} = a_0^*(Af)$, *for all $f \in \mathcal{D}(A)$.*

3) $dU([A, B]) = \overline{[dU(A), dU(B)]}$, *at least for A and B bounded and skew-self-adjoint on $\mathcal{H}$.*

Proof. This theorem combines theorem 2 and 5 in [Co]. However we give an alternative proof.

1) On an arbitrary n-particle product vector $f_1 \wedge \cdots \wedge f_n \in \mathcal{D}_A$, such that each f_j is an analytic vector for A, we have

$$
\begin{aligned}
& U(tA)a_0^*(f)U(tA)^*(f_1 \wedge \cdots \wedge f_n) \\
&= U(tA)(f \wedge e^{-tA}f_1 \wedge \cdots \wedge e^{-tA}f_n) \\
&= e^{tA}f \wedge f_1 \wedge \cdots \wedge f_n \\
&= a_0^*(e^{tA}f)(f_1 \wedge \cdots \wedge f_n)
\end{aligned}
$$

hence, $U(tA)a_0^*(f)U(-tA) = a_0^*(e^{tA}f)$ on $\mathcal{F}_\wedge(\mathcal{H})$, and the adjoint relation

$$U(tA)a_0(f)U(-tA) = a_0(e^{tA}f)$$

on $\mathcal{F}_\wedge(\mathcal{H})$, for all skew-self-adjoint operators A on $\mathcal{H}$, every analytic vectors f for A in $\mathcal{H}$ and $t \in \mathbf{R}$.

2) For each analytic vector f for A, it follows from 1) that

$$
\begin{aligned}
[dU(A), a_0^*(f)] &= \left.\frac{d}{dt}\right|_{t=0} e^{t \cdot dU(A)}a_0^*(f)e^{-t \cdot dU(A)} \\
&= \left.\frac{d}{dt}\right|_{t=0} a_0^*(e^{tA}f) \\
&= a_0^*(Af)
\end{aligned}
$$

on the domain $\mathcal{D}_A$ of $dU(A)$, giving the desired formula for all skew-self-adjoint operators A on $\mathcal{H}$ and for all analytic vectors f for A, dense in $\mathcal{D}(A)$, hence, the formula follows for all $f \in \mathcal{D}(A)$.

3) We give two different proofs for this part, the first being the more elegant one. For $f \in \mathcal{H}$ and A and B bounded and skew-self-adjoint, we have

$$
\begin{aligned}
& [dU([A, B]), a_0^*(f)] \\
&= a_0^*([A, B]f) \\
&= [dU(A), a_0^*(Bf)] - [dU(B), a_0^*(Af)] \\
&= [dU(A), [dU(B), a_0^*(f)]] - [dU(B), [dU(A), a_0^*(f)]] \\
&= [dU(A), dU(B)]a_0^*(f) - a_0^*(f)[dU(A), dU(B)] \\
&= [[dU(A), dU(B)], a_0^*(f)]
\end{aligned}
$$

on $\mathcal{D}$. Hence, the irreducibility of the Fock representation of the CAR-algebra, by theorem 1, and the fact that Ω is cancelled by $dU(\cdot)$ gives the desired formula.

We now turn to the second proof of part 3), this proof gives us some information not only on the Lie algebra level, but also on the group level. Let $t, s \in \mathbf{R}$. Since

$$e^{tA}e^{sB}e^{-tA} = \sum_{n=0}^{\infty} \frac{s^n}{n!} e^{tA}B^n e^{-tA}$$

$$= \sum_{n=0}^{\infty} \frac{s^n}{n!} (e^{tA}Be^{-tA})^n$$

$$= e^{s \cdot (e^{tA}Be^{-tA})}$$

we have

$$U(tA)U(sB)U(tA)^* a_0(f)(U(tA)U(sB)U(tA)^*)^*$$
$$= U(tA)U(sB)a_0(e^{-tA}f)U(sB)^*U(tA)^*$$
$$= a_0(e^{tA}e^{sB}e^{-tA}f)$$
$$= a_0(e^{s \cdot (e^{tA}Be^{-tA})}f)$$
$$= U(s \cdot e^{tA}Be^{-tA})a_0(f)U(s \cdot e^{tA}Be^{-tA})^*$$

Since the Fock representation of the CAR-algebra is irreducible, by theorem 1, and $U(tA)U(sB)U(tA)^*$ and $U(e^{tA}s \cdot Be^{-tA})$ are unitary, it follows that

$$U(tA)U(sB)U(tA)^* = c(t, s) \cdot U(e^{tA}sBe^{-tA})$$

where $|c(t, s)| = 1$. But, since $U(tC)\Omega = \Omega$, for any skew-self-adjoint operator C and all $t \in \mathbf{R}$, it follows that $c(t, s) = 1$. So, in fact, we have $U(tA)U(sB)U(tA)^* = U(e^{tA}s \cdot Be^{-tA})$, and for any $F \in \mathcal{D}$ we then obtain

$$[dU(A), dU(B)]F$$
$$= \frac{d}{dt}\Big|_{t=0} \frac{d}{ds}\Big|_{s=0} U(tA)U(sB)U(tA)^*F$$
$$= \frac{d}{dt}\Big|_{t=0} U(tA)dU(B)U(tA)^*F$$
$$= \frac{d}{dt}\Big|_{t=0} \frac{d}{ds}\Big|_{s=0} U(e^{tA}s \cdot Be^{-tA})F$$

$$= \left.\frac{d}{dt}\right|_{t=0} dU(e^{tA}Be^{-tA})F$$
$$= dU([A,B])F$$

from which the desired formula follows, once more. Alternatively we could easily prove 2) by direct calculation first on product vectors using equation (2.4), and then extending by linearity to all of $\mathcal{D}$. In fact, we will use this method in the case of the CCR-algebra which is treated in chapter 3   □

Later on, in section 2.4, we will use this result, but only for bounded operators A and B. Notice that theorem 2, part 3) implies that the mapping $A \to dU(A)$ is a Lie algebra homomorphism. The second quantization is sometimes called the Fock-Cook quantization mapping or just the Cook quantization mapping.

A standard example arises for $A = iI$, whereby

$$dU(A)(f_1 \wedge \cdots \wedge f_n) = i \cdot n \cdot (f_1 \wedge \cdots \wedge f_n)$$

we call $-idU(A)$ for the number operator N in $\mathcal{F}_\wedge(\mathcal{H})$, with dense domain $\mathcal{D}(N) = \{F = \oplus_{n=0}^\infty F_n \in \mathcal{F}_\wedge(\mathcal{H}) : \sum_{n=0}^\infty n^2 \cdot \|F_n\|^2 < \infty\}$. So $N(\oplus_{n=0}^\infty F_n) = \oplus_{n=0}^\infty (n \cdot F_n)$ for each $\oplus_{n=0}^\infty F_n \in \mathcal{D}(N)$. Since N is given in its spectral representation, it is evidently selfadjoint (by von Neumann's theorem, see for example [R-S 1, p. 275]), as it should be, since $dU(A)$ is skew-self-adjoint.

As a historical remark, we notice that Cook only considered second quatization in the Fock representation, this was generalized by Araki and Wyss to C^*-algebras. In 1964 Araki and Wyss in [A-W] showed that $dU(A)$ belongs to the CAR-algebra if A is a trace-class, skew-self-adjoint operator on $\mathcal{H}$, and that $dU(A)$, for finite rank operators $A : f \to Af = \sum_{i=1}^n \alpha_i \langle e_i, f \rangle e_i$, is given by $dU(A) = \sum_{i=1}^n \alpha_i a(e_i)^* a(e_i)$, where $\{e_i\}_{i=1}^n$ is an orthonormal set in $\mathcal{H}$ (most authors consider selfadjoint operators instead of skew-self-adjoint operators, we have translated their statements, as we have translated some theorems, such as Stone's and Nelson's theorem, used above). In [Lu 1], 1976, Lundberg extends the second quantization mapping, in a quasifree representation labelled by $T, 0 \leq T \leq I$, to all bounded skew-self-adjoint operators A such that $\text{Tr}(T^{\frac{1}{2}}A(I-T)AT^{\frac{1}{2}}) < \infty$, with $dU(A)$ affiliated with the CAR-algebra. In [Ar], 1985, and [Lu 2], 1990, Araki and Lundberg generalized the concept of second quantization, in the sense that they lift operators in the restricted orthogonal group to skew-self-adjoint operators acting in the Fock Hilbert spaces. We turn to this discussion in the following section. This idea was first used for the restricted unitary group, by Lundberg in [Lu 1], 1976. We point out that our formalism is different from that used in [Ar], a question we return to in section 2.6. Finally, we mention that Langmann recently has used the idea of quasi-free second quantization to obtain a current algebra in (3+1)-dimensional quantum field theory, with the well-known Mickelsson-Rajeev Schwinger term (see [Mi]), for futher details, see [La].

2.4   The Infinite–Dimensional Spin Representation

In this section we construct the so-called spin representation of the restricted orthogonal group on a Lie algebra level, by ideas similar to those of second quantization. This dicussion is closely related to the answer of the question, stated in the end of section 2.2; for which orthogonal operators T is the automorphism, given by $\pi(f) \to \pi_T(f) = \pi(T^{-1}f)$, of the Clifford algebra, unitary implementable (in the Fock representation)? In this section we use results from the former sections 2.1 – 2.3.

Many authors have studied the spin representation, we mention only a few here, [Ar], [Lu 2], [P-S] and [S-S]. We emphasize that A.M. Vershik in [V-Z] from 1990 deals with some of the same problems and in the same formalism as we do. But Vershik considers the infinite-dimensional restricted orthogonal (which he calls metagonal, in analogy with the metaplectic case), Lie algebra and group as the inductive limits of the corresponding finite dimensional ones, in his treatment of the infinite-dimensional spin group.

Any real linear mapping $T : \mathcal{H} \to \mathcal{H}$ can be split into a sum of a complex linear mapping T_1 and a complex anti-linear mapping T_2 as $T = T_1 + T_2$. If $\{e_k\}_{k \in \mathbf{N}}$ is an orthonormal basis for the complex Hilbert space $\mathcal{H}$ with inner product $\langle \cdot, \cdot \rangle$ and we let u_k and v_k denote e_k and ie_k, respectively, then $\{u_k, v_k\}_{k \in \mathbf{N}}$ form an orthonormal system with respect to $\tau(\cdot, \cdot) = Re \langle \cdot, \cdot \rangle$. We call the real span of $\{u_k, v_k\}_{k \in \mathbf{N}}$ for *the real Hilbert space*, $\mathcal{H}_r$, of $\mathcal{H}$, with inner product $\tau(\cdot, \cdot)$. Notice that $\mathcal{H}_r$ and $\mathcal{H}$ represent the same set. The complex structure on $\mathcal{H}$, given by multiplication by the imaginary unit i is reflected in $\mathcal{H}_r$ by a (real) linear bounded operator J, given by $Ju_k = v_k$ and $Jv_k = -u_k$. Then $J^2 = -I$ and $\tau(Jf, g) = -\tau(f, Jg)$, for all $f, g \in \mathcal{H}_r$, i.e. $J^\tau = -J$, where J^τ denotes the transpose of J relative to $\tau(\cdot, \cdot)$. We say that J introduces a *complex structure* in $\mathcal{H}_r$. Because of the unique correspondence between $\mathcal{H}$ and $\mathcal{H}_r$, given by $e_k \leftrightarrow u_k$ and $ie_k \leftrightarrow v_k$, we will not emphasize on which space the operators apply and therefore we omit the r-index, unless confusion may arise.

Let $T_1 = \frac{1}{2}(T - JTJ)$ and $T_2 = \frac{1}{2}(T + JTJ)$, then direct computations give that T_1 and J commute, $T_1 J = JT_1$, and that T_2 and J anti-commute, $T_2 J = -JT_2$, which gives a precise mathematical meaning to the statements that T_1 is complex linear and that T_2 is complex anti-linear. Thus, we have constructed a complex linear operator T_1 and a complex anti-linear operator T_2 such that $T = T_1 + T_2$. The subscripts 1 and 2 will in the following refer to this splitting. Consider the orthogonal group $\mathcal{O}(\mathcal{H})$, defined in the end of section 2.2, and let $T \in \mathcal{O}(\mathcal{H})$, with the above splitting $T = T_1 + T_2$. The adjoint $T_1{}^*$ of T_1 is the usual adjoint of a complex linear operator given by $\langle f, T_1 g \rangle = \langle T_1{}^* f, g \rangle$, for all $f, g \in \mathcal{H}$. But the adjoint, which we also denote by an asterisk as a superscript, $T_2{}^*$ of T_2 is given by $\langle f, T_2 g \rangle = \langle g, T_2{}^* f \rangle$, for all $f, g \in \mathcal{H}$, due to the fact that T_2 is a complex linear mapping $\mathcal{H} \to \mathcal{H}^*$, where $\mathcal{H}^*$ denotes the conjugated Hilbert space of $\mathcal{H}$, for futher details see section 2.6 later on. We restate that the adjoint operation, denoted by an asterisk, means different things, corresponding to the subscript of the operator it is applied to. Since $T^\tau T$ is the identity on $\mathcal{H}_r$, for any $T \in \mathcal{O}(\mathcal{H})$, we get $T^{-1} = T^\tau$. Since

$$
\begin{aligned}
\tau(T^\tau f, g) &= \tau(f, Tg) = \tau(f, T_1 g) + \tau(f, T_2 g) \\
&= \tau(T_1{}^* f, g) + \tau(g, T_2{}^* f) = \tau(T_1{}^* f, g) + \tau(T_2{}^* f, g) \\
&= \tau((T_1{}^* + T_2{}^*) f, g)
\end{aligned}
$$

for all $f, g \in \mathcal{H}$, where we have used the explicit form of $\tau(\cdot, \cdot)$ and its symmetry, it follows that

$$
T^{-1} = T^\tau = T_1{}^* + T_2{}^*
$$

So $T_1{}^* = (T^\tau)_1$ and $T_2{}^* = (T^\tau)_2$. Moreover

$$
\begin{aligned}
I &= T^{-1}T = T^\tau T = (T_1{}^* + T_2{}^*)(T_1 + T_2) \\
&= (T_1{}^* T_1 + T_2{}^* T_2) + (T_1{}^* T_2 + T_2{}^* T_1)
\end{aligned}
$$

so we obtain

$$T_1{}^*T_1 + T_2{}^*T_2 = I \qquad (2.5)$$

and

$$T_1{}^*T_2 + T_2{}^*T_1 = 0 \qquad (2.6)$$

since the left side of (2.5) is complex linear and the left side of (2.6) is complex anti-linear. Hence, from this and an analogous computation, it follows that T_1 and T_2 fulfil the criterion of being Bogoliubov transformations. For later use, we now define *the restricted orthogonal group* $\mathcal{O}_2(\mathcal{H})$ as the subgroup of $\mathcal{O}(\mathcal{H})$ given by

$$\mathcal{O}_2(\mathcal{H}) = \{T \in \mathcal{O}(\mathcal{H}) : T_2 \in \mathbf{L}_2(\mathcal{H})\}$$

where $\mathbf{L}_2(\mathcal{H})$ denotes the Hilbert-Schmidt operators on $\mathcal{H}$. From above it follows that $\mathcal{O}_2(\mathcal{H})$ is a group indeed. Some authors denote $\mathcal{O}_2(\mathcal{H})$ by $\mathcal{O}_J(\mathcal{H})$, due to the fact that it consists of the subgroup of $\mathcal{O}(\mathcal{H})$ such that $[T, J] = -2JT_2 \in \mathbf{L}_2(\mathcal{H})$, see for example [V-Z], who call the operators in $\mathcal{O}_2(\mathcal{H})$ *almost linear operators*.

The group $\mathcal{O}_2(\mathcal{H})$ can be given the structure of a topological group in several different ways, which is typical for infinite-dimensional groups. The strongest topology is given by the uniform topology on the complex linear part and the Hilbert-Schmidt topology on the complex anti-linear part. However, in some applications one has to use a weaker topology on the linear part, for example the strong topology (see [Ar] paragraph 6). In [Ar, p. 77] and in [P-S, p. 245] it is shown that in both these topologies, $\mathcal{O}_2(\mathcal{H})$ has two connected components, each of which is simply connected. For example, in the strongest topology it follows easily that

$$\|T^{-1}\|_{\mathcal{O}_2} = \|T_1{}^*\| + \|T_2{}^*\|_{HS} = \|T\|_{\mathcal{O}_2} < \infty$$

where $\|T\|_{\mathcal{O}_2} = \|T_1\| + \|T_2\|_{HS}$, by definition, and

$$\|ST\|_{\mathcal{O}_2} \leq \|S_1\| \cdot \|T_1\| + \|S_2\|_{HS} \cdot \|T_2\|_{HS}$$
$$+ \|S_1\| \cdot \|T_2\|_{HS} + \|S_2\|_{HS} \cdot \|T_1\| < \infty$$

since $\|\cdot\| \leq \|\cdot\|_{HS}$, where the *HS*-index on the norm symbol means the Hilbert-Schmidt norm, and

$$\|ST - HK\|_{\mathcal{O}_2}$$
$$\leq \|S_1\| \cdot \|T_1 - K_1\| + \|S_1 - H_1\| \cdot \|K_1\|$$
$$+ \|S_2\|_{HS} \cdot \|T_2 - K_2\|_{HS} + \|S_2 - H_2\|_{HS} \cdot \|K_2\|_{HS}$$
$$+ \|S_1\| \cdot \|T_2 - K_2\|_{HS} + \|S_1 - H_1\| \cdot \|K_2\|_{HS}$$
$$+ \|S_2\|_{HS} \cdot \|T_1 - K_1\| + \|S_2 - H_2\|_{HS} \cdot \|K_1\|$$

and that

$$\begin{aligned}
\|T^{-1} - S^{-1}\|_{\mathcal{O}_2} &= \|T_1^* - S_1^*\| + \|T_2^* - S_2^*\|_{HS} \\
&= \|T_1 - S_1\| + \|T_2 - S_2\|_{HS} = \|T - S\|_{\mathcal{O}_2}
\end{aligned}$$

proving that $\mathcal{O}_2(\mathcal{H})$ in fact is a topological group in the case of uniform topology on the linear part and Hilbert-Schmidt topology on the anti-linear part.

The choice of topology on $\mathcal{O}_2(\mathcal{H})$ determines the Lie algebra of $\mathcal{O}_2(\mathcal{H})$. Our choice of "pre-Lie-algebra" $o_2(\mathcal{H})$ is

$$o_2(\mathcal{H}) = \{A \in \mathbf{L}_r(\mathcal{H}) : A^\tau = -A, A_2 \in \mathbf{L}_2(\mathcal{H})\}$$

where $\mathbf{L}_r(\mathcal{H})$ denote the real linear bounded operators on $\mathcal{H}$. The phrase "pre-Lie-algebra" means that we in some applications have to enlarge the "pre-Lie-algebra" to allow operators with unbounded linear part (for futher details see [Ar, p. 81 and 104]). The demand $A^\tau = -A$ means that $A_1^* = -A_1$ and $A_2^* = -A_2$, i.e. both the linear and the anti-linear part of A are skew-selfadjoint (in their respective senses). In what follows we shall in particular consider $\mathcal{O}_2(\mathcal{H})$ in a neighbourhood of the identity, generated from $o_2(\mathcal{H})$ by the exponential mapping. Notice that the exponential mapping from an infinite-dimensional Lie algebra to the corresponding infinite dimesional Lie group, modelled on a general topological vector space, need not be locally ono-to-one nor locally onto, for example in the case for exp : $Vect(S^1) \to Diff(S^1)$, [P-S, p. 28]. In cases where the vector space is a Banach space there is a well-developed theory, which is quite parrallel to the theory of finite dimensional Lie groups.

We shall now return, as promised, to the question stated in the end of section 2.2; for which $T \in \mathcal{O}(\mathcal{H})$ is the automorphism, defined by $\pi(f) \to \pi_T(f) = \pi(T^{-1}f)$, of the Clifford algebra, unitary implementable in the Fock representation? In fact this question has already been answered by Shale and Stinespring in [S-S], 1965, as stated in the following theorem.

Theorem 3 *A unitary operator $U(T)$, which implements the automorphism $\pi(f) \to \pi_T(f)$ exists if and only if $T \in \mathcal{O}_2(\mathcal{H})$. Moreover, the operator $U(T)$ is unique up to a phase of modulus one.*

Proof. A proof can be found in [S-S], however, we make a construction of $U(T)$ below, in a neighbourhood of the identity in $\mathcal{O}_2(\mathcal{H})$. □

Because the unique correspondance between the a's and the π's the irreducibility of the C^*-algebra generated by $\{a_0(f) : f \in \mathcal{H}\}$ implies the irreducibility of the C^*-algebra generated by $\{\pi(f) : f \in \mathcal{H}\}$ as well. But

$$U(TS)^{-1}\pi(f)U(TS) = (U(T)U(S))^{-1}\pi(f)(U(T)U(S))$$

so $U(T)U(S)U(TS)^{-1}$ equals a constant times the identity I. Hence

$$U(T)U(S) = c(T,S) \cdot U(TS)$$

where $c(T,S) \in \mathbf{C}$. Now, the unitarity of $U(\cdot)$ forces $c(T,S)$ to be of modulus one

$$
\begin{aligned}
\langle f, g \rangle &= \langle U(T)U(S)f, U(T)U(S)g \rangle \\
&= |c(T,S)|^2 \cdot \langle U(TS)f, U(TS)g \rangle \\
&= |c(T,S)|^2 \cdot \langle f, g \rangle
\end{aligned}
$$

for all $f, g \in \mathcal{H}$, giving that $|c(T,S)| = 1$. This means that the mapping $T \to U(T)$ is a projective representation of the restricted orthogonal group $\mathcal{O}_2(\mathcal{H})$.

The group cocycle $c(T,S)$ depends on the choice of the arbitrary phase in $U(T)$. It is possible to give an explicit formula for the cocycle $c(T,S)$, by choosing $U(T)$ such that $c(T,S)$ is smooth in such a way that $U(\cdot)$ lift one-parameter groups into one-parameter groups, for T and S close to the identity. We do this below, by giving a constructive proof of the if-part of the above theorem by Shale and Stinespring in the case of T in a neighbourhood of the identity in $\mathcal{O}_2(\mathcal{H})$ consisting of elements of the form $T = e^A$, with $A \in o_2(\mathcal{H})$. This is done by constructing the spin representation, on a Lie algebra level, that is, we construct $U(e^{sA})$ for $A \in o_2(\mathcal{H})$ and $s \in \mathbf{R}$, by constructing its skew-selfadjoint generator $dU(A)$, hence $U(e^{sA})$ is given by $e^{s \cdot dU(A)}$.

Consider first the complex linear part A_1 of $A \in o_2(\mathcal{H})$, which is skew-selfadjoint. In this case $dU(A_1)$ and $U(e^{sA_1})$ are constructed by the method of Cook's second quantization given in section 2.3, where we denoted $U(e^{sA_1})$ by $U(s \cdot A_1)$. Hence, $dU(A_1)$ is skew-selfadjoint on $\mathcal{D}$, where $\mathcal{D}$ is the dense set of analytic vectors for $dU(A_1)$, given by vectors $F = \oplus_{n=0}^{\infty} F_n \in \mathcal{F}_\wedge(\mathcal{H})$ obeying that only finitely many F_n are non-zero, due to the boundedness of A and consequently of A_1. Observe that $dU(A_1) : \wedge^n \mathcal{H} \to \wedge^n \mathcal{H}$.

We now turn to the anti-linear part A_2 of $A \in o_2(\mathcal{H})$. Since A_2 is a Hilbert-Schmidt operator, there exist two orthogonal sets $\{u_i\}_{i \in I}$ and $\{v_i\}_{i \in I}$ in $\mathcal{H}$, both spanning the range of A_2, such that A_2 has the representation

$$A_2 f = \sum_{i \in I} \langle f, v_i \rangle u_i$$

for any $f \in \mathcal{H}$. Notice that A_2 given in this form clearly is anti-linear, due to our convention that the inner product is anti-linear in the first argument. A direct calculation, using the definition of the adjoint of an anti-linear operator and the skew-selfadjointness of A_2, yields

$$\sum_{i \in I} \langle f, v_i \rangle u_i = -\sum_{i \in I} \langle f, u_i \rangle v_i$$

Moreover, $\|A_2\|_{HS}^2 = \sum_{i \in \mathbb{N}} \|A_2 e_i\|^2 = \sum_{i \in I} \|v_i\|^2 \cdot \|u_i\|^2$. This means that we may identify A_2 with a vector $\mathcal{A}_2 \in \wedge^2 \mathcal{H}$, where

$$\mathcal{A}_2 = \sum_{i \in I} v_i \wedge u_i = \sqrt{2} \cdot \sum_{i \in I} v_i \otimes u_i$$

Observe that the mapping $A_2 \mapsto \mathcal{A}_2$ is not an isometry, but the mapping $A_2 \mapsto \frac{1}{\sqrt{2}} \mathcal{A}_2$ is. We are now able to map A_2 into an operator on $\mathcal{F}_\wedge(\mathcal{H})$ by generalizing the idea of the creation operator. For the rest of this chapter, we discard the 0-index on $a_0(f)$, since we shall only consider the Fock representation. Define the operator $a^*(\mathcal{A}_2)$ by

$$a^*(\mathcal{A}_2)\Omega = \mathcal{A}_2 \tag{2.7}$$

and

$$a^*(\mathcal{A}_2)(f_1 \wedge \cdots \wedge f_n) = \mathcal{A}_2 \wedge f_1 \wedge \cdots \wedge f_n$$

and extend by linearity to the domain $\mathcal{D}$. To prove that $a^*(\mathcal{A}_2)$ is well defined as an operator from $\wedge^n \mathcal{H}$ into $\wedge^{n+2} \mathcal{H}$ we first notice that if $I = I_N$ is finite then $a^*(\mathcal{A}_2) = \sum_{i \in I_N} a^*(v_i) a^*(u_i)$. This expression does not depend on the representation $\mathcal{A}_2 = \sum_{i \in I_N} v_i \wedge u_i$. In fact, if $\mathcal{A}_2 = \sum_{i \in I_N} v'_i \wedge u'_i$ is another representation, choose an orthonormal basis $\{e_k\}_{k \in I_N}$ for $\text{span}\{v_i\}$ $(= \text{span}\{v'_i\} = \text{span}\{u_i\} = \text{span}\{u'_i\})$, hence

$$a^*(\mathcal{A}_2) = \sum_{i,k,l \in I_N} \langle e_k, v_i \rangle \langle e_l, u_i \rangle a^*(e_k) a^*(e_l)$$

is identical to

$$a^*(\mathcal{A}_2) = \sum_{i,k,l \in I_N} \langle e_k, v'_i \rangle \langle e_l, u'_i \rangle a^*(e_k) a^*(e_l)$$

since

$$A = \sum_{i,k,l \in I_N} \sqrt{2} \langle e_k, v_i \rangle \langle e_l, u_i \rangle e_k \otimes e_l = \sum_{i,k,l \in I_N} \sqrt{2} \langle e_k, v_i' \rangle \langle e_l, u_i' \rangle e_k \otimes e_l$$

and $\{e_k \otimes e_l\}$ form an orthonormal basis for $\otimes^2 \mathcal{H}$.

Now, we show that $a^*(A_2) : \wedge^n \mathcal{H} \to \wedge^{n+2} \mathcal{H}$ is bounded by $\sqrt{(n+2)(n+1)} \cdot \|A\|$. As above we first consider the case when $I = I_N$ is finite. Let

$$F_n = \sum_{s \in M_1} f_{s,3} \wedge \cdots \wedge f_{s,n+2} \in \wedge^n \mathcal{H}$$

and

$$G_{n+2} = \sum_{t \in M_2} g_{t,1} \wedge \cdots \wedge g_{t,n+2} \in \wedge^{n+2} \mathcal{H}$$

where M_1 and M_2 are finite subsets of $\mathbf{N}$, and let $\{e_k\}_{k \in I_M}$ be a finite orthogonal basis for

$$\text{span}\{v_i, f_{s,3}, \ldots, f_{s,n+2}, g_{t,1} \ldots, g_{t,n+2} : i \in I_N, s \in M_1, t \in M_2\}$$

Then

$$A_2 = \sum_{i_1,i_2 \in I_M} \alpha(i_1, i_2) \cdot e_{i_1} \otimes e_{i_2} \in \wedge^2 \mathcal{H}$$

$$F_n = \sum_{i_3,\ldots,i_{n+2} \in I_M} \beta(i_3, \ldots, i_{n+2}) \cdot e_{i_3} \otimes \cdots \otimes e_{i_{n+2}} \in \wedge^n \mathcal{H}$$

$$G_{n+2} = \sum_{i_1,\ldots,i_{n+2} \in I_M} \gamma(i_1, \ldots, i_{n+2}) \cdot e_{i_1} \otimes \cdots \otimes e_{i_{n+2}} \in \wedge^{n+2} \mathcal{H}$$

with α, β and γ totally antisymmetric in their arguments (especially $\alpha(i_1, i_2) = 0$ for i_1 or i_2 in $I_M \setminus I_N$.) Hence, by the Cauchy-Schwarz's inequality

$$\langle a^*(A_2) F_n, G_{n+2} \rangle = \sqrt{\frac{(n+2)(n+1)}{2}} \cdot \sum_{i_1,\ldots,i_{n+2} \in I_M} \overline{\alpha \cdot \beta} \cdot \gamma$$

$$\leq \sqrt{\frac{(n+2)(n+1)}{2}}$$
$$\cdot \left(\sum_{i_1,i_2 \in I_M} |\alpha|^2 \right)^{\frac{1}{2}} \left(\sum_{i_3,\ldots,i_{n+2} \in I_M} |\beta|^2 \right)^{\frac{1}{2}} \left(\sum_{i_1,\ldots,i_{n+2} \in I_M} |\gamma|^2 \right)^{\frac{1}{2}}$$
$$\leq \sqrt{(n+2)(n+1)} \cdot \|A_2\| \cdot \|F_n\| \cdot \|G_{n+2}\|$$

where $\alpha = \alpha(i_1, i_2)$, $\beta = \beta(i_3, \ldots, i_{n+2})$ and $\gamma = \gamma(i_1, \ldots, i_{n+2})$. Since this enequality holds for all G_{n+2} of the above form, giving a dense subspace of $\wedge^{n+2}\mathcal{H}$, it follows that

$$\|a^*(A_2)F_n\| \leq \sqrt{(n+2)(n+1)} \cdot \|A_2\| \cdot \|F_n\|$$

for all F_n of the above form, hence for all $F_n \in \wedge^n\mathcal{H}$. Consider now the case when I is not necessarily finite. It follows from above that $a^*(A_2)$ is well-defined, with

$$\begin{aligned}
a^*(A_2)(f_1 \wedge \cdots \wedge f_n) &= A_2 \wedge f_1 \wedge \cdots \wedge f_n \\
&= \lim_{N \to \infty} (A_2^{(N)} \wedge f_1 \wedge \cdots \wedge f_n)
\end{aligned}$$

where $A_2 = \lim_{N \to \infty} A_2^{(N)}$ and $A_2^{(N)} = \sum_{i=1}^{N} v_i \wedge u_i \in \wedge^2\mathcal{H}$. Thus

$$\|a^*(A_2)F_n\| \leq \sqrt{(n+2)(n+1)} \cdot \|A_2\| \cdot \|F_n\|$$

for any $F_n \in \wedge^n\mathcal{H}$. Hence, the formula holds for any $F \in \mathcal{D}$, since if $F_n = 0$ for $n > N$ then each term $\sqrt{(n+2)(n+1)}$ is dominated by $\sqrt{(N+2)(N+1)}$ and we get the same formula but with N instead of n. So $a^*(A_2)$ is well-defined on $\mathcal{D}$, which we may take as the domain. Notice that $a^*(A_2)$ is in general unbounded, since if $\{e_{i_1}, \ldots, e_{i_n}\}_{i_1 < \cdots < i_n}$ is an orthonormal system orthogonal to span$\{v_i\}_{i \in I}$, which equals the range of A_2, then

$$\begin{aligned}
\|a^*(A_2)(e_{i_1} \wedge \cdots \wedge e_{i_n})\| &= \|A_2 \wedge e_{i_1} \wedge \cdots \wedge e_{i_n}\| \\
&= \sqrt{(n+2)(n+1)} \cdot \|A_2\|
\end{aligned}$$

which clearly is unbounded, due to the n-dependent factor. The domain $\mathcal{D}$ is evidently invariant under the action of $a^*(A_2)$, so $\mathcal{D}$ forms a invariant dense set for $a^*(A_2)$. In fact, $\mathcal{D}$ is a dense set of analytic vectors for $a^*(A_2)$, since

$$\|a^*(A_2)^k F\| \leq \sqrt{\frac{(N+2k)!}{N!}} \cdot \|A_2\|^k \cdot \|F\|$$

for any $F \in \mathcal{D}$ such that $F_n = 0$ for $n > N$, so

$$\sum_{k=0}^{\infty} \frac{|t|^k}{k!} \|a^*(A_2)^k F\| \leq \sum_{k=0}^{\infty} \frac{c(k)}{k!} (|t| \|A_2\|)^k \cdot \|F\|$$

which is finite for $|t| < (2\|A_2\|)^{-1}$ and $F \in \mathcal{D}$, with $F_n = 0$ for $n > N$, by use of the ratio test, where $c(k) = \sqrt{\frac{(N+2k)!}{N!}}$.

Notice that

$$
\begin{aligned}
A_2 f = \sum_{i\in I} \langle f, v_i \rangle u_i &= \frac{1}{2}\sum_{i\in I} a(f)(v_i \wedge u_i) \\
&= \frac{1}{2}a(f)A_2 = \frac{1}{2}a(f)a^*(A_2)\Omega
\end{aligned}
$$

so $a^*(A_2 f)\Omega = \frac{1}{2}[a(f), a^*(A_2)]\Omega$, for all $f \in \mathcal{H}$. In fact this commutation relation holds, not only on Ω, but on all of $\mathcal{D}$. Let $F_n = f_1 \wedge \cdots \wedge f_n$, then

$$
\begin{aligned}
a(f)a^*(A_2)F_n &= a(f)(A_2 \wedge F_n) \\
&= (a(f)A_2) \wedge F_n + A_2 \wedge (a(f)F_n) \\
&= (2a^*(A_2 f) + a^*(A_2)a(f))F_n
\end{aligned}
$$

where we have used the invariance of $\mathcal{D}$ under the action of $a(f)$, hence

$$
\frac{1}{2}[a(f), a^*(A_2)] = a^*(A_2 f) \tag{2.8}
$$

on $\mathcal{D}$, for all $f \in \mathcal{H}$. Observe that $a^*(A_2) : \wedge^n \mathcal{H} \to \wedge^{n+2}\mathcal{H}$. By $a(A_2)$ we denote the formal adjoint $a^*(A_2)^*$ of $a^*(A_2)$. It follows that it is well-defined on $\mathcal{D}$. Since $\langle \Omega, a^*(A_2)F \rangle = 0$, for all $F \in \mathcal{D}$, we have that $\Omega \in \mathcal{D}(a(A_2))$ and $a(A_2)\Omega = 0$. On $\wedge^1 \mathcal{H}$ it follows analogously that $f \in \mathcal{D}(a(A_2))$ and $a(A_2)f = 0$, for each $f \in \mathcal{H}$. We continue by induction after $n \in \mathbf{N}$ using the fact that $a^*(A_2)$ increase the number of particle by two and using formula (2.8). Suppose that we have proved that $f_2 \wedge \cdots \wedge f_n \in \mathcal{D}(a(A_2))$ for any $f_2, \ldots, f_n \in \mathcal{H}$, then, for any $f_1 \in \mathcal{H}$ and $F \in \wedge^{n-2}\mathcal{H}$

$$
\begin{aligned}
\langle f_1 \wedge \cdots \wedge f_n, a^*(A_2)F \rangle \\
= \langle a^*(f_1)(f_2 \wedge \cdots \wedge f_n), a^*(A_2)F \rangle \\
= \langle f_2 \wedge \cdots \wedge f_n, a(f_1)a^*(A_2)F \rangle \\
= \langle f_2 \wedge \cdots \wedge f_n, (a^*(A_2)a(f_1) + 2a^*(A_2 f_1))F \rangle \\
= \langle (a^*(f_1)a(A_2) + 2a(A_2 f_1))(f_2 \wedge \cdots \wedge f_n), F \rangle
\end{aligned}
$$

which is well-defined, since $f_2 \wedge \cdots \wedge f_n \in \mathcal{D}(a(A_2))$, i.e.

$$
f_1 \wedge \cdots \wedge f_n \in \mathcal{D}(a(A_2))
$$

and

$$a(A_2)(f_1 \wedge \cdots \wedge f_n) = (a^*(f_1)a(A_2) + 2a(A_2 f_1))(f_2 \wedge \cdots \wedge f_n)$$

Hence, by linearity, it follows that $\mathcal{D} \subset \mathcal{D}(a(A_2))$. Now let $G = \oplus_{n=0}^{\infty} G_n \in \mathcal{D}$, such that $G_n = 0$, for $n > N$, and let $F = \oplus_{n=0}^{\infty} F_n \in \mathcal{D}$. Then

$$|\langle G, a^*(A_2)F \rangle| \leq \sqrt{N(N-1)} \cdot \|A_2\| \cdot \|G\| \cdot \|F\|$$

where we have used the estimate

$$\|a^*(A_2)F_n\| \leq \sqrt{(n+2)(n+1)} \|A_2\| \cdot \|F_n\|$$

for $n \leq N - 2$. So

$$\|a(A_2)G\| = \sup_{F \in \mathcal{D}, \|F\|=1} |\langle a(A_2)G, F \rangle| \leq \sqrt{N(N-1)} \cdot \|A_2\| \cdot \|G\|$$

since $\mathcal{D}$ is dense in $\mathcal{H}$. Moreover, we obtain

$$\|a(A_2)^k G\| \leq C(k) \cdot \|A_2\|^k \cdot \|G\|$$

where $C(k) = 0$, for $2k > N$, and $C(k) = \sqrt{\frac{N!}{(N-2k)!}}$, for $2k \leq N$. Then

$$\sum_{k=0}^{\infty} \frac{|t|^k}{k!} \|a(A_2)^k G\| \leq \sum_{k=0}^{\infty} \frac{(|t| \cdot \|A_2\|)^k}{k!} \cdot C(k) \cdot \|G\|$$

is finite, for $|t| < (2 \cdot \|A_2\|)^{-1}$, due to the ratio test, since the quotients of the k'th and the $(k-1)$'th term converge to $2 \cdot |t| \cdot \|A_2\|$, as $k \to \infty$, which is stricly less that one, for $|t| < (2 \cdot \|A_2\|)^{-1}$. Hence, each vector in $\mathcal{D}$ is an analytic vector for $a(A_2)$. Observe that $a(A_2) : \wedge^n \mathcal{H} \to \wedge^{n-2} \mathcal{H}$, yields zero for $n = 0, 1$. Since $\mathcal{D}$ is invariant under the action of $a^*(f)$, the adjoint of formula (2.8) also holds on $\mathcal{D}$, $\frac{1}{2}[a(A_2), a^*(f)] = a(A_2 f)$, for all $f \in \mathcal{H}$. So $\mathcal{D}$ is invariant under the action of both $a^*(A_2)$ and $a(A_2)$.

Define $dU(A_2) = \frac{1}{2}(a(A_2) - a^*(A_2))$ on $\mathcal{D}$, for A_2 a skew-selfadjoint anti-linear Hilbert-Schmidt operator. Notice that $dU(A_2) : \wedge^n \mathcal{H} \to \wedge^{n-2} \mathcal{H} \oplus \wedge^{n+2} \mathcal{H}$. Define $dU(A) = dU(A_1) + dU(A_2)$ on $\mathcal{D}$, for $A = A_1 + A_2 \in o_2(\mathcal{H})$. Observe that $dU(A) : \wedge^n \mathcal{H} \to \wedge^{n-2} \mathcal{H} \oplus \wedge^n \mathcal{H} \oplus \wedge^{n+2} \mathcal{H}$.

Theorem 4 *The operator $dU(A)$ is essentially skew-selfadjoint and fulfils*

$$[dU(A), \pi(f)] = \pi(Af) \qquad (2.9)$$

on $\mathcal{D}$, for all $f \in \mathcal{H}$ and all $A \in o_2(\mathcal{H})$,

$$\langle \Omega, dU(A)\Omega \rangle = 0 \qquad (2.10)$$

for all $A \in o_2(\mathcal{H})$, and

$$\langle \Omega, dU(A)dU(B)\,\Omega \rangle = -\frac{1}{4}\langle A_2, B_2 \rangle = \frac{1}{2}\mathrm{Tr}(B_2 A_2) \qquad (2.11)$$

Proof. The operator $dU(A)$ is skew-symmetric and has $\mathcal{D}$ as a dense set of analytic vectors (it is not in general true that $\mathcal{D}'$ is a set of analytic vectors for the sum of two operators which both have $\mathcal{D}'$ as a set of analytic vectors, however, in this particular case one may benefit from the fact that $\mathcal{D}$ is a dense set of analytic vectors for respectively $a^*(A_2)$, $a(A_2)$ and $dU(A_1)$). Then it follows by a modification of Nelson's theorem ([R-S 2, p. 202]) to skew-symmetric operators that $dU(A)$ is essential skew-selfadjoint. From theorem 2 we have that $[dU(A_1), \pi(f)] = \pi(A_1 f)$, on $\mathcal{D}$ for all $f \in \mathcal{H}$ (since it holds for π replaced by a and a^*, respectively). For the anti-linear part of A we get

$$
\begin{aligned}
[dU(A_2), \pi(f)] &= \frac{1}{2}[a(A_2), \pi(f)] - \frac{1}{2}[a^*(A_2), \pi(f)] \\
&= 2^{-3/2}[a(A_2), a(f)^*] + 2^{-3/2}[a(A_2), a(f)] \\
&\quad - 2^{-3/2}[a^*(A_2), a(f)^*] - 2^{-3/2}[a^*(A_2), a(f)] \\
&= 2^{-1/2}a(A_2 f) + 0 - 0 + 2^{-1/2}a^*(A_2 f) \\
&= \pi(A_2 f)
\end{aligned}
$$

on $\mathcal{D}$, for all $f \in \mathcal{H}$, where we have used that $\pi(f) = 2^{-1/2}(a(f) + a(f)^*)$, formula (2.8) and its adjoint, and that $[a^*(A_2), a(f)^*] = 0$ together with the adjoint relation on $\mathcal{D}$. Hence

$$
\begin{aligned}
[dU(A), \pi(f)] &= [dU(A_1), \pi(f)] + [dU(A_2), \pi(f)] \\
&= \pi(A_1 f) + \pi(A_2 f) = \pi(Af)
\end{aligned}
$$

on $\mathcal{D}$, for all $f \in \mathcal{H}$, proving (2.9). Moreover, since $dU(A_1)\Omega = 0$ and $a(A_2)\Omega = 0$, it follows that

$$\langle \Omega, dU(A)\Omega \rangle = -\frac{1}{2}\langle \Omega, a^*(A_2)\Omega \rangle = 0$$

proving (2.10). Finally

$$\begin{aligned}
\langle \Omega, dU(A)dU(B)\,\Omega \rangle &= -\langle dU(A)\Omega, dU(B)\Omega \rangle \\
&= -\langle \frac{1}{2}a^*(A_2)\Omega, \frac{1}{2}a^*(B_2)\Omega \rangle \\
&= -\frac{1}{4}\langle A_2, B_2 \rangle \\
&= \frac{1}{2}\mathrm{Tr}(B_2 A_2)
\end{aligned}$$

on $\mathcal{D}$, where we have used the anti-symmetry of $dU(A)$, definition (2.7) and, in the last equality, the spectral forms of A_2 and B_2, respectively, combined with the definition of the trace, proving (2.11). This will be explicitly done in the symmetric case (see theorem 17).

$\square$

We now define the unitary one-parameter group $U(e^{sA})$ as

$$U(e^{sA}) = e^{s \cdot dU(A)}$$

At this point we are rather close to having proved the if part of theorem 3 in a neighbourhood of the identity, but the fact that $U(e^{sA})$ create infinite many particles, is the reason why we first have to prove the following technical lemma, which doesn't seem to appear anywhere else.

Lemma 5 *Let G be an essential skew-selfadjoint operator with $\mathcal{D}$ as a dense set of analytical vectors and let B be any bounded operator leaving $\mathcal{D}$ invariant, both defined on the same Hilbert space. Then*

$$e^{sG}Be^{-sG} = \sum_{n=0}^{\infty} \frac{s^n}{n!}[G, B]^{(n)}$$

on $\mathcal{D}$ for all $s \geq 0$, where G also denotes the closure of the corresponding essential skew-selfadjoint operator, $[G, B]^{(0)} = B$, $[G, B]^{(1)} = [G, B]$ and $[G, B]^{(n)} = [G, [G, B]^{(n-1)}]$, inductively, for $n \in \mathbf{N}$.

Proof. Let $f \in \mathcal{D}$, then $e^{-sG}f$ is well-defined, and so is $Be^{-sG}f$. Let $g \in \mathcal{D}$, then $\langle e^{-sG}g, Be^{-sG}f \rangle$ is well-defined, and induction after $n \in \mathbf{N}$ gives

$$\frac{d^n}{ds^n}\langle e^{-sG}g, Be^{-sG}f \rangle = \langle e^{-sG}g, [G, B]^{(n)}e^{-sG}f \rangle$$

so, by Taylors formula, which is valid for $f, g \in \mathcal{D}$, we get

$$\langle e^{-sG}g, Be^{-sG}f \rangle = \sum_{n=0}^{\infty} \frac{s^n}{n!} \frac{d^n}{ds^n}\Big|_{s=0} \langle e^{-sG}g, Be^{-sG}f \rangle$$

$$= \langle g, \sum_{n=0}^{\infty} \frac{s^n}{n!}[G, B]^{(n)}f \rangle$$

for all $f, g \in \mathcal{D}$, which is dense in the Hilbert space. Then $Be^{-sG}f \in \mathcal{D}((e^{-sG})^*)$ and $e^{sG}Be^{-sG} = \sum_{n=0}^{\infty} \frac{s^n}{n!}[G, B]^{(n)}$ on $\mathcal{D}$.   □

The above lemma, with $G = dU(A)$ and $B = \pi(f)$ (and the substitution $s \to -s$), gives

$$U(e^{sA})^*\pi(f)U(e^{sA}) = \sum_{n=0}^{\infty} \frac{(-s)^n}{n!}[dU(A), \pi(f)]^{(n)}$$

$$= \sum_{n=0}^{\infty} \frac{(-s)^n}{n!}\pi(A^n f)$$

$$= \pi(e^{-sA}f)$$

on $\mathcal{D}$, for all $f \in \mathcal{H}$, where we have used (2.9). Hence, we have the desired formula

$$U(T)^{-1}\pi(f)U(T) = \pi(T^{-1}f) = \pi_T(f) \qquad (2.12)$$

on $\mathcal{H}$ for all $T = e^{sA}$, $A \in o_2(\mathcal{H})$, $f \in \mathcal{H}$ and $s \in \mathbf{R}$, where $U(T)$ has been explicit constructed, such that the arbitrary phase of $U(T)$ has been fixed on all one-parameter subgroups of $\mathcal{O}_2(\mathcal{H})$ of the form $T = e^{sA}$, $A \in o_2(\mathcal{H})$. We call $U : T \to U(T)$ the *spin representation* of the restricted orthogonal group and we define the *spin group $Spin_2$* to be the group of all the unitary implementers $U(T)$, $T \in \mathcal{O}_2(\mathcal{H})$ from Shale and Stinespring's theorem (theorem 3). This construction is more transparent on the Lie algebra level.

Theorem 6 *The elements $dU(A)$, $A \in o_2(\mathcal{H})$, form a Lie algebra on $\mathcal{D}$ with bracket*

$$[dU(A), dU(B)] = dU([A, B]) + \omega(A, B) \cdot I \qquad (2.13)$$

and the Lie algebra cocycle is given by

$$\omega(A, B) = -\frac{1}{2}\mathrm{Tr}([A_2, B_2]) = -\frac{i}{2}Im\langle A_2, B_2 \rangle \qquad (2.14)$$

This infinite-dimensional Lie algebra is denoted the spin Lie algebra, $spin_2(\mathcal{H})$.

Proof. By use of (2.9) we have

$$
\begin{aligned}
[dU(A)dU(B), \pi(f)] &= dU(A)[dU(B), \pi(f)] + [dU(A), \pi(f)]dU(B) \\
&= dU(A)\pi(Bf) + \pi(Af)dU(B)
\end{aligned}
$$

on $\mathcal{D}$, for all $f \in \mathcal{H}$. So

$$
\begin{aligned}
[[dU(A), dU(B)], \pi(f)] &= dU(A)\pi(Bf) + \pi(Af)dU(B) \\
&\quad - dU(B)\pi(Af) - \pi(Bf)dU(A) \\
&= [dU(A), \pi(Bf)] + [\pi(Af), dU(B)] \\
&= \pi(ABf) - \pi(BAf) \\
&= \pi([A, B]f) \\
&= [dU([A, B]), \pi(f)]
\end{aligned}
$$

on $\mathcal{D}$, for all $f \in \mathcal{H}$, again by (2.9). Then the irreducibility of $\{\pi(f) : f \in \mathcal{H}\}$ implies

$$
[dU(A), dU(B)] = dU([A, B]) + \omega(A, B) \cdot I
$$

The equations (2.10) and (2.11) give us an explicit formula for the Lie algebra cocycle

$$
\begin{aligned}
\omega(A, B) &= \langle \Omega, \omega(A, B) \cdot I \Omega \rangle \\
&= \langle \Omega, [dU(A), dU(B)]\Omega \rangle \\
&= -\frac{1}{2}\mathrm{Tr}([A_2, B_2]) \\
&= -\frac{i}{2}Im\langle A_2, B_2 \rangle
\end{aligned}
$$

where we have used (2.10),(2.11), and (2.13).   □

Notice that the trace of a commutator of two anti-linear Hilbert-Schmidt operators do not vanish, in general, as in the complex linear case. Moreover, it follows directly from (2.13) that ω is skew-symmetric and fulfils the Hochshild condition

$$
\omega(AB, C) + \omega(BC, A) + \omega(CA, B) = 0
$$

for all $A, B, C \in o_2(\mathcal{H})$, since $[AB, C] + [BC, A] + [CA, B] = 0$ for all bounded operators A, B and C. Thus, also the Jacobi identity is fulfilled

$$\omega([A, B], C) + \omega([B, C], A) + \omega([C, A], B) = 0$$

for all $A, B, C \in o_2(\mathcal{H})$, hence ω is a closed two-form.

Corollary 7 *The mapping $A \to dU(A)$, $A \in o_2(\mathcal{H})$, is a projective representation, from the Lie algebra $o_2(\mathcal{H})$ onto the spin Lie algebra $spin_2(\mathcal{H})$, with cocycle given by (2.14).*

Proof. Follows directly from theorem 6.   □

The cocycle given by (2.14) is studied in detail in [V-Z], article 1, paragraph 1.2, and it defines a non-trivial central extension of the Lie algebra $o_2(\mathcal{H})$. In the special case where the linear part of A and B are trace-class operators, we are able to transform the cocycle term away, by a change of phase, as follows: Put $dU_0(A) = dU(A) - \frac{1}{2} \cdot \mathrm{Tr}(A_1) \cdot I$, then a straight forward calculation gives $[dU_0(A), dU_0(B)] = dU_0([A, B])$, by (2.14) and the fact that $\frac{1}{2} \cdot \mathrm{Tr}([A, B]_1) = -\omega(A, B)$. So we put $U_0(e^{sA}) = e^{s \cdot dU_0(A)} = e^{-\frac{1}{2}s\mathrm{Tr}(A_1)} U(e^{sA})$, for $s \in \mathbf{R}$ close to zero. Then

$$U(e^{sC}) = e^{\frac{1}{2}s(\mathrm{Tr}(C_1) - \mathrm{Tr}(A_1) - \mathrm{Tr}(B_1))} U(e^{sA}) U(e^{sB})$$

where C is given explicitly by the Campbell-Baker-Hausdorff formula, such that $e^{sC} = e^{sA} e^{sB}$, for s close to zero, i.e. the group cocycle $c(e^{sA}, e^{sB})$ is given by

$$c(e^{sA}, e^{sB}) = (\det(e^{sA_1} e^{sB_1} e^{-sC_1}))^{\frac{1}{2}}$$

for s close to zero, where $\det(e^D) = e^{\mathrm{Tr}(D)}$, for any trace class operators D.

We finish this section by calculating an explicit formula for *the vacuum functional*, given by $c(s) = \langle \Omega, U(e^{sA})\Omega \rangle$, for $A \in o_2(\mathcal{H})$ and s in a neighbourhood of zero. Notice that $c(s)$ is analytic at $s = 0$, since Ω is an analytic vector for the generator $dU(A)$.

Put $T = e^{sA}$ and consider $\Omega_s = U(e^{sA})\Omega$. Then, by (2.12), we have $(a(T_1 f) + a^*(T_2 f))\Omega_s = U(T)a(f)\Omega = 0$, for all $f \in \mathcal{H}$. Define the anti-linear Hilbert-Schmidt operator K by $K = T_2 T_1^{-1}$, where $T = T_1 + T_2$, T_2 is an anti-linear Hilbert-Schmidt operator and

T_1 is linear and invertible for s sufficiently small, by the Neumann series. Observe that K is skew-selfadjoint $\langle f, Kg \rangle = \langle f, T_2 T_1^{-1} g \rangle = \langle T_1^{-1} g, T_2^* f \rangle = \langle g, (T_1^{-1})^* T_2^* f \rangle = \langle g, -T_2 T_1^{-1} f \rangle = \langle g, -Kf \rangle$, for all $f, g \in \mathcal{H}$, where we have used that $(T_1^{-1})^* T_2^* = -T_2 T_1^{-1}$, by formula (2.6) applied to $S = T_1^{-1} + T_2^*$. It then follows from the above that $(a(g) + a^*(Kg))\Omega_s = 0$, for all $g \in \mathcal{H}$ $(f = T_1^{-1} g)$.

Now, we prove that

$$\Omega_s = c(s) \cdot e^{-\frac{1}{2} a^*(K)} \Omega$$

Put $\Omega_s = \oplus_{n=0}^{\infty} \Omega_n(s)$, where $\Omega_n(s) \in \wedge^n \mathcal{H}$. Note that $\Omega_0(s) = c(s) \cdot \Omega$. It follows from the above that

$$
\begin{aligned}
0 &= (a(g) + a^*(Kg))\Omega_s \\
&= a(g)\Omega_1(s) \oplus (\oplus_{n=1}^{\infty}(a(g)\Omega_{n+1}(s) + a^*(Kg)\Omega_{n-1}(s)))
\end{aligned}
$$

So $\Omega_1(s) = 0$, since $a(g)\Omega_1(s) = 0$, for all $g \in \mathcal{H}$, then by induction it follows that $\Omega_{2n-1}(s) = 0$, for $n \in \mathbf{N}$. Moreover, since $a(g)\Omega_0(s) = 0$ and formula (2.8) holds, it follows that

$$a(g)(\Omega_2(s) + \frac{1}{2}a^*(K)\Omega_0(s)) = a(g)\Omega_2(s) + a^*(Kg)\Omega_0(s) = 0$$

for all $g \in \mathcal{H}$, hence $\Omega_2(s) = -\frac{1}{2}a^*(K)\Omega_0(s) = -\frac{1}{2}c(s)K$. Now, induction after $n \in \mathbf{N}$ yields $\Omega_{2n}(s) = \frac{1}{n!}(-\frac{1}{2}a^*(K))^n \Omega_0(s)$. For $n = 1$ we have already proved the formula. Let $n \in \mathbf{N}$ and suppose that $\Omega_{2n}(s) = \frac{1}{n!}(-\frac{1}{2}a^*(K))^n \Omega_0(s)$, then

$$
\begin{aligned}
a^*(K)a(g)a^*(K)^n \Omega &= 2n\, a^*(K)a^*(Kg)a^*(K)^{n-1}\Omega \\
&= 2n\, a^*(Kg)a^*(K)^n \Omega
\end{aligned}
$$

where the first equality is a result of the spectral resolution of K and the second equality is a consequence of $[a^*(Kg), a^*(K)] = 0$. So

$$
\begin{aligned}
n \cdot a^*(Kg)\Omega_{2n}(s) &= \frac{1}{2}a^*(K)a(g)\Omega_{2n}(s) \\
&= (\frac{1}{2}a(g)a^*(K) - a^*(Kg))\Omega_{2n}(s)
\end{aligned}
$$

by formula (2.8), and

$$a^*(Kg)\Omega_{2n}(s) = \frac{1}{n+1} \cdot \frac{1}{2}a(g)a^*(K)\Omega_{2n}(s)$$

Hence

$$0 = a(g)\Omega_{2(n+1)}(s) + a^*(Kg)\Omega_{2n}(s)$$

$$= a(g)(\Omega_{2(n+1)}(s) + \frac{1}{n+1} \cdot \frac{1}{2}a^*(K)\Omega_{2n}(s))$$

for all $g \in \mathcal{H}$, from which it follows that

$$\Omega_{2(n+1)}(s) = \frac{1}{n+1}(-\frac{1}{2}a^*(K))\,\Omega_{2n}(s)$$

$$= \frac{1}{(n+1)!}(-\frac{1}{2}a^*(K))^{n+1}\Omega_0(s)$$

proving the desired formula for $n+1$. Thus

$$\Omega_s = \oplus_{n=0}^{\infty}\frac{1}{n!}(-\frac{1}{2}a^*(K))^n\Omega_0(s)$$

$$= e^{-\frac{1}{2}a^*(K)}\Omega_0(s)$$

$$= c(s)e^{-\frac{1}{2}a^*(K)}\Omega$$

since $\Omega_0(s) = c(s) \cdot \Omega$. This formula allow us to get a differential equation for $c(s)$, as follows

$$c'(s) = \frac{d}{ds}\langle\Omega, \Omega_s\rangle = \langle\Omega, dU(A)\Omega_s\rangle$$

$$= -\langle dU(A)\Omega, \Omega_s\rangle = \frac{1}{2}\langle a^*(A_2)\Omega, \Omega_s\rangle$$

$$= \frac{1}{2}c(s)\langle A_2, e^{-\frac{1}{2}a^*(K)}\Omega\rangle = -\frac{1}{4}c(s)\langle A_2, K\rangle$$

$$= \frac{1}{2}\mathrm{Tr}(KA_2) \cdot c(s)$$

where we have used formula (2.11), the just derived formula for Ω_s above and that $dU(A)\Omega = -\frac{1}{2}a^*(A_2)\Omega$. So $c(s)$ is given by the above differential equation and the fact that $c(0) = \|\Omega\|^2 = 1$. Notice that $K = T_2T_1^{-1} = (e^{sA})_2(e^{sA})_1^{-1}$ depends on $s \in \mathbf{R}$. Put $V_s = e^{-sA_1}(e^{sA})_1$, then

$$\frac{d}{ds}V_s = e^{-sA_1}(-A_1)(e^{sA})_1 + e^{-sA_1}(Ae^{sA})_1$$

$$= e^{-sA_1}(-A_1)(e^{sA})_1 + e^{-sA_1}(A_1(e^{sA})_1 + A_2(e^{sA})_2)$$

$$= e^{-sA_1}A_2(e^{sA})_2$$

and for s sufficiently small, so that V_s is close to the identity I and therefore invertible, we obtain

$$
\begin{aligned}
V_s^{-1} \frac{d}{ds} V_s &= (e^{sA})_1^{-1} e^{sA_1} e^{-sA_1} A_2 (e^{sA})_2 \\
&= (e^{sA})_1^{-1} A_2 (e^{sA})_2 = T_1^{-1} A_2 T_2
\end{aligned}
$$

Because $V_s - I = \int_0^s e^{-tA_1} A_2 (e^{tA})_2 \, dt$ is a trace-class operator, since A_2 and $(e^{tA})_2$ both are Hilbert-Schmidt operators, $\log V_s$ does exist. Hence

$$
\begin{aligned}
\frac{d}{ds} \operatorname{Tr}(\log V_s) &= \operatorname{Tr}(V_s^{-1} \frac{d}{ds} V_s) = \operatorname{Tr}(T_1^{-1} A_2 T_2) \\
&= \operatorname{Tr}(A_2 T_2 T_1^{-1}) = \operatorname{Tr}(A_2 K) = \overline{\operatorname{Tr}(K A_2)}
\end{aligned}
$$

where we have used that V_s^{-1} and $\frac{d}{ds} V_s$ commute (only) under the trace symbol, since both are complex linear operators, as is the case for T_1^{-1} and $A_2 T_2$, and that $\overline{\operatorname{Tr}(K A_2)} = \operatorname{Tr}(A_2 K)$, due to $\langle e_j, K A_2 e_j \rangle = \langle e_j, A_2 K e_j \rangle$, for an arbitrary basis vector $e_j \in \mathcal{H}$. Then we may write the differential equation as

$$
c'(s) = \frac{1}{2} \frac{d}{ds} \overline{\operatorname{Tr}(\log(V_s))} \cdot c(s)
$$

which has the solution

$$
c(s) = \kappa \cdot e^{\frac{1}{2} \overline{\operatorname{Tr}(\log(V_s))}} = \kappa \cdot (\overline{\det(V_s)})^{\frac{1}{2}} = \kappa \cdot (\det(V_{-s}))^{\frac{1}{2}}
$$

since $c(-s) = \overline{c(s)}$, and we have used that the determinant of V_s exists, due to the fact that $V_s - I$ is a trace class operator. Finally, it follows that $\kappa = 1$, since $c(0) = 1$, so

$$
c(s) = (\det(V_{-s}))^{\frac{1}{2}}
$$

giving an explicit formula for $c(s)$, as claimed. We summarize the above in the following theorem.

Theorem 8 *The vacuum functional $c(s) = \langle \Omega, U(e^{sA}) \Omega \rangle$ for $A \in o_2(\mathcal{H})$ and s in a neighbourhood of zero, where $U(\cdot)$ denotes the spin representation, is simply*

$$
c(s) = (\det(V_{-s}))^{\frac{1}{2}}
$$

where $V_{-s} = e^{sA_1} (e^{-sA})_1 = I - \int_0^s e^{tA_1} A_2 (e^{-tA})_2 \, dt$.

Proof. An immediate consequence of the above.   □

Since the (restricted) unitary group can be realized as a subgroup of the (restriced) orthogonal group, we may study the restriction of the spin representation to the restricted unitary group. We do this in the following section.

2.5   The Spin Representation of the Restricted Unitary Group

As mentioned at the end of the previous section, we will now consider the restriction of the spin representation to the restricted unitary group, partly because we get a nice explicit expression for the Lie algebra cocycle, which we will use later on in an application of the theory on a loop group. This is allowed by the fact that we may realize the restricted unitary group as a subgroup of the restricted orthogonal group.

Let P be an orthogonal projection on the Hilbert space $\mathcal{H}$ and let $\mathcal{U}(\mathcal{H})$ denote the unitary group on $\mathcal{H}$. We define the restricted unitary group $\mathcal{U}_2(\mathcal{H}, P)$ on $\mathcal{H}$ by

$$\mathcal{U}_2(\mathcal{H}, P) = \{V \in \mathcal{U}(\mathcal{H}) : [P, V] \in \mathbf{L}_2(\mathcal{H})\}$$

where $\mathbf{L}_2(\mathcal{H})$, as earlier, denotes the Hilbert-Schmidt operators on $\mathcal{H}$. We wish to say a few things about the corresponding "pre-Lie-algebra", which we denote $u_2(\mathcal{H}, P)$. We demand that the elements of $u_2(\mathcal{H}, P)$, by the exponential mapping, define unitary one-parameter groups in $\mathcal{U}_2(\mathcal{H}, P)$, in resemblance with the preceding sections. Suppose $A \in u_2(\mathcal{H}, P)$, and consider $V_s = e^{sA}$. From the unitarity of V_s it follows, by taking the s-derivative at $s = 0$, that A is skew-selfadjoint, $A^* = -A$ (and that A is complex linear). Of course, it would be convenient if $[P, V_s] \in \mathbf{L}_2(\mathcal{H})$, where $V_s = e^{sA}$, implies that $[P, A] \in \mathbf{L}_2(\mathcal{H})$, this is, however, not the case, since $\mathbf{L}_2(\mathcal{H})$ is closed only with respect to the Hilbert-Schmidt topology, and not the uniform topology. We define

$$u_2(\mathcal{H}, P) = \{A \in \mathbf{L}(\mathcal{H}) : A^* = -A, [P, A] \in \mathbf{L}_2(\mathcal{H})\}$$

where $\mathbf{L}(\mathcal{H})$ denotes the bounded linear operators on $\mathcal{H}$, later on we may want to enlarge our choice of "pre-Lie-algebra" to unbounded operators.

It is evident that e^{sA} defines a unitary operator on $\mathcal{H}$, for $A \in u_2(\mathcal{H}, P)$, so the non-trivial part, in proving that $e^{sA} \in \mathcal{U}_2(\mathcal{H}, P)$, is to prove that $[P, e^{sA}]$ is Hilbert-Schmidt, for any $A \in u_2(\mathcal{H}, P)$. Split A into two parts

$$B = (I - P)A(I - P) + PAP$$

and

$$C = PA(I - P) + (I - P)AP$$

such that $A = B + C$. It follows that both B and C are skew-self-adjoint and that C is Hilbert-Schmidt, since $PA(I - P) = P[P, A]$ and $[P, A]$ is Hilbert-Schmidt by assumption. Moreover $[P, B] = 0$. Define $V(s) = e^{-sB}e^{sA}$, $s \in \mathbf{R}$, which is a unitary operator. We now show that $V(s) - I$ may be written as a series which converges in the Hilbert-Schmidt topology, uniformly in s, on compact sets in $\mathbf{R}$. Observe that

$$\frac{d}{ds}V(s) = e^{-sB}(A - B)e^{sA}$$

$$= e^{-sB}Ce^{sA}$$

$$= C(s)V(s)$$

where $C(s) = e^{-sB}Ce^{sB}$. Integration then gives

$$V(s) = I + \int_0^s C(t)V(t)\, dt$$

since $V(0) = I$. Put $V_0(s) = I$ and iterate the equation by putting

$$V_{n+1}(s) = \int_0^s C(t)V_n(t)\, dt$$

for $n \in \mathbf{N} \cup \{0\}$. Then

$$\|V_1(s)\|_{HS} \leq \|C\|_{HS} \cdot \left|\int_0^s dt\,\right| = \|C\|_{HS} \cdot |s|$$

and

$$\|V_{n+1}(s)\|_{HS} = \left\|\int_0^s C(t)V_n(t)\, dt\right\|_{HS}$$

$$\leq \left|\int_0^s \|C(t)V_n(t)\|_{HS} dt\,\right|$$

$$\leq \left|\int_0^s \|C(t)\|_{HS}\|V_n(t)\|_{HS} dt\,\right|$$

$$= \|C\|_{HS} \cdot \left|\int_0^s \|V_n(t)\|_{HS} dt\,\right|$$

for $n \in \mathbb{N}$, since $\|C(t)\|_{HS} = \|C\|_{HS}$. Assuming that $\|V_n(s)\|_{HS} \le \|C\|_{HS}^n \frac{|s|^n}{n!}$, then

$$
\begin{aligned}
\|V_{n+1}(s)\|_{HS} &\le \|C\|_{HS} \cdot | \int_0^s \|V_n(t)\|_{HS} dt | \\
&\le \|C\|_{HS}^{n+1} \cdot \frac{1}{n!} \cdot | \int_0^s |t|^n dt | \\
&= \|C\|_{HS}^{n+1} \cdot \frac{|s|^{n+1}}{(n+1)!}
\end{aligned}
$$

Thus

$$
\| \sum_{n=1}^{\infty} V_n(s)\|_{HS} \le \sum_{n=1}^{\infty} \|V_n(s)\|_{HS} \le e^{\|C\|_{HS} \cdot |s|}
$$

So the series $\sum_{n=1}^{\infty} V_n(s)$ converges in the Hilbert-Schmidt topology, uniformly in s on compact sets in $\mathbb{R}$. Moreover, $\sum_{n=0}^{\infty} V_n(s)$ is a solution to the integral equation, determining $V(s)$. Hence $V(s) - I = \sum_1^{\infty} V_n(s)$ is Hilbert-Schmidt.

Since $[P, B] = 0$, it follows that

$$
P e^{sA}(I - P) = P e^{sB} V(s)(I - P) = e^{sB} P(V(s) - I)(I - P)
$$

is Hilbert-Schmidt, because

$$
\begin{aligned}
\|P e^{sA}(I - P)\|_{HS} &= \|e^{sB} P(V(s) - I)(I - P)\|_{HS} \\
&= \|P(V(s) - I)(I - P)\|_{HS} \\
&\le \|V(s) - I\|_{HS} \\
&< \infty
\end{aligned}
$$

for s in a compact subset of $\mathbb{R}$. Hence,

$$
[P, e^{sA}] = P e^{sA}(I - \dot{P}) - (I - P)\dot{e^{sA}} P
$$

is Hilbert-Schmidt, so $e^{sA} \in \mathcal{U}_2(\mathcal{H}, P)$.

Now, let Γ be an involution on $\mathcal{H}$ commuting with P (as in section 2.2 such anti-unitary involutions do exist). Put $I_P = I - P + \Gamma P$ and observe that $I_P^2 = (I - P)^2 + (\Gamma P)^2 = I - P + \Gamma^2 P = I$, so the real linear operator I_P is invertible on $\mathcal{H}$. Moereover

$$
\begin{aligned}
\tau(I_P f, I_P g) &= \tau((I - P)f, g) + \tau(\Gamma P f, \Gamma P g) \\
&= \tau((I - P)f, g) + \tau(P g, P f) \\
&= \tau(f, g)
\end{aligned}
$$

for all $f, g \in \mathcal{H}$, where $\tau(\cdot, \cdot)$, as earlier, denotes $Re(\cdot, \cdot)$. This means that $I_P \in \mathcal{O}(\mathcal{H})$. By use of this, the restricted unitary group $\mathcal{U}_2(\mathcal{H}, P)$ can be realized as a subgroup of the restricted orthogonal group $\mathcal{O}_2(\mathcal{H})$. For $V \in \mathcal{U}_2(\mathcal{H}, P)$ we put $V_P = I_P V I_P$, then $V \to V_P$ defines a representation of $\mathcal{U}_2(\mathcal{H}, P)$ in $\mathcal{O}_2(\mathcal{H})$, since V_P evidently is real linear and invertible, in fact, $V_P^{-1} = I_P V^* I_P$, and a direct calculation shows that $\tau(V_P f, V_P g) = \tau(f, g)$, using the unitarity of V and that $I_P \in \mathcal{O}(\mathcal{H})$, so $V_P \in \mathcal{O}(\mathcal{H})$. Finally

$$
\begin{aligned}
V_P &= I_P V I_P \\
&= (I - P)V(I - P) + P\Gamma V\Gamma P + (I - P)V\Gamma P + \Gamma P V(I - P) \\
&= (I - P)V(I - P) + P\Gamma V\Gamma P + \Gamma P[P, V] - [P, V]\Gamma P \\
&= (V_P)_1 + (V_P)_2
\end{aligned}
$$

where $(V_P)_1 = (I - P)V(I - P) + P\Gamma V\Gamma P$ is the linear part of V and the anti-linear part $(V_P)_2 = \Gamma P[P, V] - [P, V]\Gamma P = [\Gamma P, [P, V]]$ is Hilbert-Schmidt, due to the fact that $[P, V] \in \mathbf{L}_2(\mathcal{H})$, since $V \in \mathcal{U}_2(\mathcal{H}, P)$. Hence $V_P \in \mathcal{O}_2(\mathcal{H})$.

This allows us to construct the spin representation of the restricted unitary group, and the corresponding subgroup of $Spin_2(\mathcal{H})$ will be denoted $Spin_2(\mathcal{H}, P)$.

Define $U_P(V) = U(V_P)$, for any $V \in \mathcal{U}_2(\mathcal{H}, P)$, where $U(\cdot)$ is the spin representation, defined in section 2.4; $U(V_P)$ is well-defined, since $V_P \in \mathcal{O}_2(\mathcal{H})$. Moreover, we put $a_P(f) = a((I-P)f) + a^*(\Gamma P f)$, for $f \in \mathcal{H}$, which clearly is anti-linear, and $f \to a_P(f)$ evidently gives a representation of the CAR-algebra labelled by P, since $(I - P)$ and ΓP are Bogoliubov transformations (see section 2.2, just above theorem 1). Then

$$
\pi_{I_P}(f) = \pi(I_P f) = \frac{1}{\sqrt{2}}(a_P(f) + a_P{}^*(f))
$$

for $f \in \mathcal{H}$, gives a representation of the Clifford algebra, where we have used that $I_P{}^{-1} = I_P$. By use of formula (2.12) and the fact that $I_P V = V_P I_P$ it follows that

$$
\begin{aligned}
\pi_{I_P}(Vf) &= \pi(I_P V f) = \pi(V_P I_P f) \\
&= U(V_P)\pi(I_P f)U(V_P)^{-1} \\
&= U_P(V)\pi_{I_P}(f)U_P(V)^{-1}
\end{aligned}
$$

and consequently

$$
a_P(Vf) = U_P(V)a_P(f)U_P(V)^{-1}
$$

for $f \in \mathcal{H}$ and $V \in \mathcal{U}_2(\mathcal{H}, P)$. Of course, we also have the analogous formula for $a_P{}^*(\cdot)$.

Let $A \in u_2(\mathcal{H}, P)$ and define $dU_P(A)$ as the generator of the unitary one-parameter group $U_P(e^{sA}) = U(I_P e^{sA} I_P)$. In the case where A is bounded, as in the above choice of pre-Lie-algebra $u_2(\mathcal{H}, P)$, it follows that $I_P e^{sA} I_P = e^{sA_P}$ and then $dU_P(A) = dU(A_P)$. For $A = iI$, which is in $u_2(\mathcal{H}, P)$, $U_P(e^{siI}) = e^{sdU_P(iI)} = e^{sdU((iI)_P)}$ and $(iI)_P = i(I - 2P)$, which, in fact, is complex linear, so

$$dU_P(iI) = dU((i(I - 2P))_1) = idU(I - 2P)$$

where the last equality follows because the generator reduces to the linear Fock-Cook generator. Then we may define the so-called charge operator Q as the selfadjoint operator $Q = -idU_P(iI) = dU(I-2P)$, which is well-defined at least on $\mathcal{D}$. To discuss the spectrum of Q we define $\mathcal{H}_- = P\mathcal{H}$ and $\mathcal{H}_+ = (I - P)\mathcal{H} = \mathcal{H}_-^\perp$ such that $\mathcal{H} = \mathcal{H}_+ \oplus \mathcal{H}_-$ and let $\{e_j : j \in M_\pm\}$ denote an arbitrary choice of orthonormal basis for respectively $\mathcal{H}_+$ and $\mathcal{H}_-$. Notice that this splitting of $\mathcal{H}$ into $\mathcal{H}_+ \oplus \mathcal{H}_-$ equips $\mathcal{H}$ with a polarization, defined by the unitary operator $J = I - 2P$, on $\mathcal{H}$, which is $+1$ on $\mathcal{H}_+$ and -1 on $\mathcal{H}_-$, since $JP = -P$ and $J(I - P) = (I - P)$. Hence, requirement of $[P, V] \in \mathbf{L}_2(\mathcal{H})$ is equivalent to $[J, V] \in \mathbf{L}_2(\mathcal{H})$, which is the formulation used in [P-S, p. 80]. In a more physical language, J is a one-particle charge operator corresponding to the case where $\mathcal{H}_\pm$ is the space representing particle with respectively positive $(+)$ and negative $(-)$ charge, so $Je_k = q_k \cdot e_k$, where $q_k = \pm 1$ for $k \in M_\pm$, respectively, i.e. q_k is the sign of the charge of the particle described by the state vector e_k. Then $Q = dU(J)$ on $\mathcal{D}$. Now, it follows that $\Omega \in \mathcal{D}$ and $Q\Omega = 0$, by definition. Moreover, $e_{j_1} \wedge \cdots \wedge e_{j_n} \in \mathcal{D}$ and

$$Q(e_{j_1} \wedge \cdots \wedge e_{j_n}) = \sum_{k=1}^{n} e_{j_1} \wedge \cdots \wedge Je_{j_k} \wedge \cdots \wedge e_{j_n}$$

$$= \left(\sum_{k=1}^{n} q_{j_k}\right) \cdot (e_{j_1} \wedge \cdots \wedge e_{j_n})$$

notice that $\sum_{k=1}^{n} q_{j_k} \in \mathbf{Z}$. Evidently, we have

$$\|Q(e_{j_1} \wedge \cdots \wedge e_{j_n})\| \leq n \cdot \|e_{j_1} \wedge \cdots \wedge e_{j_n}\|$$

so $\mathcal{D}(N) \subset \mathcal{D}(Q)$, where N denotes the number operator, discussed in the end of section 2.3. Observe that the eigenvalue corresponding to $e_{j_1} \wedge \cdots \wedge e_{j_n}$ belongs to $\{-n, -n + 2, \ldots, n - 2, n\}$, where

$n \in \mathbf{N}$, and all possibilities can occur, of course dependent of the choice of the eigenvector. This means that the set of eigenvalues of Q at least is $\mathbf{Z}$ and that the eigenvectors corresponding to a fixed eigenvalues $q \in \mathbf{Z}$ are all in the subspace spanned by those product basis vectors, made up of anti-symmetric tensor products of one-particle vectors with q particles more from $\mathcal{H}_{sign(q)}$ than from the orthogonal complement $\mathcal{H}_{sign(q)}^{\perp} = \mathcal{H}_{-sign(q)}$ (here the sign of $q = 0$ is optional). As a consequence, there are infinitely many eigenvectors to each eigenvalue, $q \in \mathbf{Z}$. Moreover, Q is evidently unbounded, since its spectrum is unbounded. Up to now we have proved that the spectrum of Q includes $\mathbf{Z}$, we shall now prove that the spectrum of Q indeed is $\sigma(Q) = \mathbf{Z}$. For all $s \in \mathbf{R}$ is $e^{siQ} = U_P(e^{siI})$ $= U_P(e^{iI(s+2\pi)}) = e^{siQ} \cdot e^{i2\pi Q}$, then $e^{i2\pi Q} = I$, which implies that $\sigma(Q) \subset \mathbf{Z}$, hence $\sigma(Q) = \mathbf{Z}$. Moreover, for $\lambda \notin \mathbf{Z}$, the range of $Q - \lambda$ is dense in $\mathcal{F}_\wedge(\mathcal{H})$ and

$$\|(Q - \lambda \cdot I)F\| \geq c_\lambda \cdot \|F\|$$

where $c_\lambda = dist(\lambda, \mathbf{Z}) > 0$, so $(Q - \lambda)^{-1}$ is well-defined with bounded c_λ^{-1}. Notice that the spectrum consists of eigenvalues only. For each $q \in \mathbf{Z} = \sigma(Q)$, let $\mathcal{H}_q \subset \mathcal{F}_\wedge(\mathcal{H})$ denote the eigenspace of Q corresponding to the eigenvalue q. Then $\mathcal{H}_q \perp \mathcal{H}_{q'}$, for $q \neq q'$, both in $\mathbf{Z}$. Hence, we are able to decompose $\mathcal{F}_\wedge(\mathcal{H})$ in the following way

$$\mathcal{F}_\wedge(\mathcal{H}) = \oplus_{q \in \mathbf{Z}} \mathcal{H}_q$$

called the charge gradation of $\mathcal{F}_\wedge(\mathcal{H})$. If $A \in u_2(\mathcal{H}, P)$, then Q commute with $dU_P(A)$, by use of (1.13), (1.14) and the fact that $[A_P, B_P] = I_P[A, B]I_P = [A, B]_P$, for $A, B \in u_2(\mathcal{H}, P)$, since

$$
\begin{aligned}
[dU_P(A), Q] &= [dU(A_P), dU(J)] \\
&= dU([A_P, J]) + \omega(A_P, J) \cdot I \\
&= dU([A, iI]_P) - i\omega(A_P, (iI)_P) \cdot I \\
&= 0 - \frac{i}{2}\text{Tr}([(A_P)_2, (I_P iI_P)_2]) \cdot I \\
&= 0
\end{aligned}
$$

because $I_P i I_P = i(I - 2P) = iJ$ is linear. Notice that this argument only holds for A bounded. However, Q does not commute with all operators $U_P(V) = U(V_P)$, $V \in \mathcal{U}_2(\mathcal{H}, P)$ (as can be seen from the spin representation of the charge group C, which will be treated later on). The above charge gradation and the fact that Q commutes with $dU_P(A)$, and then with $U_P(e^{sA})$, for $A \in u_2(\mathcal{H}, P)$, implies that the operator $U_P(e^{sA})$ maps $\mathcal{H}_q$ into $\mathcal{H}_q$. That the operator $U_P(e^{sA})$, for $A \in u_2(\mathcal{H}, P)$, leaves $\mathcal{H}_q$ invariant, means that $U_P(e^{sA})$ conserves the charge. But, since not all operators $U_P(V)$, for $V \in \mathcal{U}_2(\mathcal{H}, P)$, commute with Q, they do not leave $\mathcal{H}_q$ invariant, and therefore they do not, in general, conserve the charge.

Let us calculate an explicit formula for the Lie algebra cocycle, in the case where $A, B \in u_2(\mathcal{H}, P)$, by use formula (2.11) in theorem 4, as follows

$$
\begin{aligned}
\langle \Omega, dU_P(A)dU_P(B)\,\Omega \rangle \\
= \ & \langle \Omega, dU(A_P)dU(B_P)\,\Omega \rangle \\
= \ & \frac{1}{2}\mathrm{Tr}((B_P)_2(A_P)_2) \\
= \ & \frac{1}{2}\mathrm{Tr}((I - P)BPA(I - P) + \Gamma PB(I - P)AP\Gamma) \\
= \ & \frac{1}{2}\mathrm{Tr}((I - P)BPA(I - P)) + \frac{1}{2}\mathrm{Tr}((PB(I - P)AP)^*) \\
= \ & \frac{1}{2}\mathrm{Tr}(PA(I - P)BP) + \frac{1}{2}\mathrm{Tr}(PA(I - P)BP) \\
= \ & \mathrm{Tr}(PA(I - P)BP) \qquad (2.15)
\end{aligned}
$$

where we have used that

$$
\begin{aligned}
(C_P)_2 \ &= \ (I_P C I_P)_2 \\
&= \ ((I - P)C(I - P) + \Gamma PC\Gamma P \\
& \qquad + (I - P)C\Gamma P + \Gamma PC(I - P))_2 \\
&= \ (I - P)C\Gamma P + \Gamma PC(I - P)
\end{aligned}
$$

for $C = A, B$, that $\Gamma^2 = I$, that $\mathrm{Tr}(\Gamma C\Gamma) = \mathrm{Tr}(C^*)$, that $A^* = -A$, $B^* = -B$, that $(I - P)BP = [B, P]P$ and that $PA(I - P) = P[P, A]$ both are linear Hilbert-Schmidt operators such that

$$
\mathrm{Tr}((I - P)BPA(I - P)) = \mathrm{Tr}(PA(I - P)BP)
$$

Hence, from formula (2.14), in theorem 6, we get the following explicit expression for the Lie algebra cocycle

$$
\begin{aligned}
\omega(A_P, B_P) \ &= \ \langle \Omega, [dU(A_P), dU(B_P)]\,\Omega \rangle \\
&= \ -\frac{1}{2}\mathrm{Tr}([(A_P)_2, (B_P)_2]) \qquad (2.16) \\
&= \ \mathrm{Tr}(PA(I - P)BP) - \mathrm{Tr}(PB(I - P)AP)
\end{aligned}
$$

so $\omega_P(A, B) = \mathrm{Tr}(PA(I - P)BP) - \mathrm{Tr}(PB(I - P)AP)$, where $\omega_P(A, B) = \omega(A_P, B_P)$ and the analogue of formula (2.13) then reads

$$[dU_P(A), dU_P(B)] = dU_P([A, B]) + \omega_P(A, B) \cdot I \qquad (2.17)$$

for $A, B \in u_2(\mathcal{H}, P)$.

We end our treatment of the spin representation, based on the CAR-algebra and its Fock representation, by the following diagram, reflecting our construction.

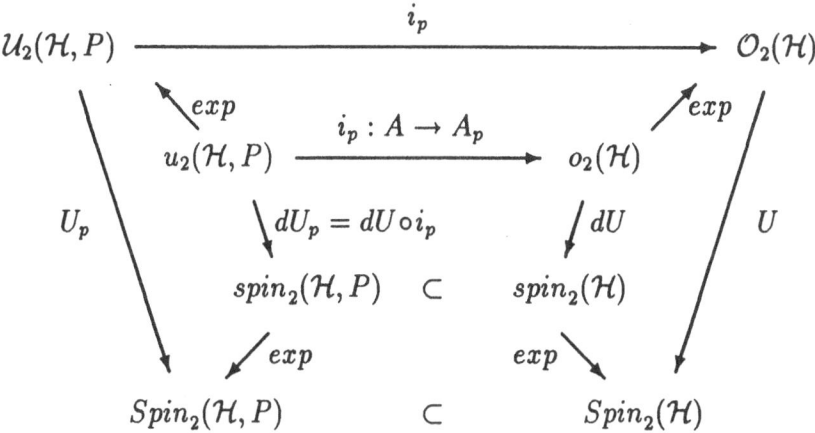

We return to some applications in a later chapter. Before doing so, we will make an analogous discussion of the CCR-algebra. However, we conclude this chapter with a section discussing different formalisms.

2.6 A Note on the Formalism

We have several times referred to other authors, using another formalism, see for example [Ar]. This section is devoted to the correspondance between our formalism, which is consistent with the formalism used in some reference papers, such as [Lu 2], [B-R 2] and

[V-Z], and the formalism used by others, such as [Ar] and [Ya]. To clarify the discussion, we will only consider the transformation between the formalism used in this book and the formalism used by Araki, in [Ar] and [Ya]. The discussion of this section doesn't seem to appear anywhere else.

Let $\mathcal{H}$ denote a complex Hilbert space and let $\mathcal{H}^*$ denote the conjugate Hilbert space, that is $\mathcal{H}^*$ denotes the same set as $\mathcal{H}$, with the same addition rule, but with the complex conjugated scalar multiplication, $\mathbf{C} \times \mathcal{H} \to \mathcal{H}$ (relative to $\mathcal{H}$) given by $\lambda \cdot f = \overline{\lambda} f$ and the complex conjugated inner product, $\mathcal{H} \times \mathcal{H} \to \mathbf{C}$ given by $\langle f, g \rangle_{\mathcal{H}^*} = \overline{\langle f, g \rangle}_{\mathcal{H}} = \langle g, f \rangle_{\mathcal{H}}$. Put $\mathcal{K} = \mathcal{H} \oplus \mathcal{H}^*$ and write an $h \in \mathcal{K}$ as $h = h_1 + h_2$, where $h_1 \in \mathcal{H}$ and $h_2 \in \mathcal{H}^*$. Then $\langle h, h' \rangle_{\mathcal{K}} = \langle h_1, h_1' \rangle_{\mathcal{H}} + \overline{\langle h_2, h_2' \rangle}_{\mathcal{H}}$. This is the Hilbert space used in [Ar].

Define a orthogonal projection P onto $\mathcal{H}$ on $\mathcal{K}$ by $Ph = h_1$, for any $h \in \mathcal{K}$, where $h = h_1 + h_2$ refers to the above splitting. We also define a complex conjugation on $\mathcal{K}$ by $\Gamma(h_1, h_2) = (h_2, h_1)$, for $h_1 \in \mathcal{H}$, $h_2 \in \mathcal{H}^*$, where we write (h_1, h_2) for $h = h_1 + h_2$. Notice that $\Gamma \lambda h = \overline{\lambda} \Gamma h$. We may write $\Gamma = \begin{pmatrix} 0 & I \\ I & 0 \end{pmatrix}$. Observe that $P + \Gamma P \Gamma = I$, hence P is what Araki calls a basis projection. Moreover, Araki defines $Re_{\Gamma}\mathcal{K} = \{h = (h_1, h_2) \in \mathcal{K} : h_1 = h_2\}$ $= \{h \in \mathcal{K} : \Gamma h = h\}$. Hence, $\langle (h, h), (h', h') \rangle_{\mathcal{K}} = 2 \cdot Re \langle h, h' \rangle_{\mathcal{H}}$, for $(h, h), (h', h') \in Re_{\Gamma}\mathcal{K}$. Then Araki defines a generalized Clifford operator (relative to ours) by

$$\begin{aligned} \pi_P(B(h)) &= a(Ph)^* + a(P\Gamma h) \\ &= a((h_1, 0))^* + a((h_2, 0)) \\ &= a(h_1)^* + a(h_2) \end{aligned}$$

in the Fock representation (here π_P denotes the Fock representation given by P). However, on $Re_{\Gamma}\mathcal{K}$, the generalized Clifford operator matches with our Clifford operator $\pi(h) = a(h)^* + a(h)$, where π denotes our Clifford operator, defined only on the space $Re_{\Gamma}\mathcal{K}$, which corresponds to our Hilbert space (which we in the preceding sections denote $\mathcal{H}$). The advantage of Araki's formalism is that orthogonal transformations correspond to unitary operators, which commute with Γ, and then one is able to use spectral theory directly. However, this formalism pays the price, in dealing with a Hilbert space twice as big as our "physical" Hilbert space, $Re_{\Gamma}\mathcal{K}$.

A (unitary) operator U on $\mathcal{K} = \mathcal{H} \oplus \mathcal{H}^*$ may be written as

$$U = \begin{pmatrix} A & B \\ C & D \end{pmatrix}$$

The requirement that $[U, \Gamma] = 0$ then gives

$$B = C = W = PU(I - P) + (I - P)UP$$

and

$$A = D = V = PUP + (I - P)U(I - P)$$

The unitarity of U then gives

$$V^*V + W^*W = I = VV^* + WW^*$$

and

$$V^*W + W^*V = 0 = WV^* + VW^*$$

which is equivalent to our decomposition of an orthogonal transformation $T = T_1 + T_2$, with $T_1 = V$ and $T_2 = W$. Observe that $T = V + W$ in fact becomes an orthogonal transformation on $Re_\Gamma \mathcal{K}$ and that $W : \mathcal{H} \oplus \mathcal{H}^* \to (\mathcal{H} \oplus \mathcal{H}^*)^*$ is complex linear, corresponding to a complex anti-linear operator on $Re_\Gamma \mathcal{K}$. Hence

$$U = \begin{pmatrix} T_1 & T_2 \\ T_2 & T_1 \end{pmatrix}$$

explains the connection of the transformation between operators used in the respective formalisms.

Chapter 3

The Metaplectic Representation of the Infinite-Dimensional Symplectic Group

3.1 The Canonical Commutation Relations and the CCR-Algebra

In this section we introduce the canonical commutation relations and the CCR-algebra. Although the CCR-algebra has many properties analogous to the CAR-algebra treated in Chap. 2, there is a lack of norm continuity in the definition of the CCR-algebra. This lack of norm continuity is related to the fact that creation and annihilation operators in any representation are unbounded. Physically, this reflects the main difference between fermions, related to the CAR-algebra, obeying the Pauli principle (see Chap. 2), and bosons, related to the CCR-algebra, which do not obey the Pauli principle. So in the case of bosons there is no bound on the number of particles which can occupy a given physical state. This is reflected by the unboundedness of the annihilation and creation operators mentioned above. Usually one treats the boson case from a slightly different viewpoint than the fermion case due to the previously mentioned qualitative difference between the two cases. One uses the so-called Weyl operators which, roughly speaking, are the unitary operators constructed from the self-adjoint closure of the sum of the annihilation and creation operators, to study a bosonic many-particle system. The Weyl operators fulfil the so-called Weyl form of the canonical commutation relations.

The canonical commutation relations were introduced by Dirac (see [Dir]) in 1927 in the context of radiation theory in physics. However, Cook was probably the first to give a rigorous mathematical approach to the subject (see [Co]). It must be mentioned that the abstract structure of creation and annihilation operators has been studied in the 1950's. In particular, Segal (see [Se,I.E.]) emphasized the C^*-algebraic structure of the CCR-algebra and introduced severel new features and results.

In the following sections we will use the specific realization of the canonical commutation relations, called the Fock representation of the canonical commutation relations. We therefore focus on this representation. We include, however, a general discussion of the CCR-algebra serving as a natural frame similar to the case of the CAR-algebra.

This section is based on Chap. 2 in this book, [Co] and [B-R 2]. We will use a notation similar to that of Chap. 2, but common symbols do in general have different meaning.

Let $\mathcal{H}$ be a complex Hilbert space and $\mathcal{A}$ an abstract $*$-algebra. We call $\mathcal{A}$ the $*$-algebra generated by $\{a(f) : f \in \mathcal{H}\}$, if there is an anti-linear mapping $a : \mathcal{H} \mapsto \mathcal{A}$ such that $\{a(f) : f \in \mathcal{H}\}$ together whith the unit I generate the $*$-algebra $\mathcal{A}$ and fulfil *the canonical commutation relations*

$$\begin{aligned}
[a(f), a(g)] &= 0 \\
[a(f), a(g)^*] &= \langle f, g \rangle \cdot I
\end{aligned} \tag{3.1}$$

for all $f, g \in \mathcal{H}$. Here $\langle \cdot, \cdot \rangle$ denotes the inner product on $\mathcal{H}$ and $[\cdot, \cdot]$ the usual commutator $[A, B] = AB - BA$. We will in brief refer to the canonical commutation relations given in (3.1) as the CCR.

Assume that we have a C^*-norm on $\mathcal{A}$, i.e. a norm fulfilling that $\|A^*A\| = \|A\|^2$ for all $A \in \mathcal{A}$. The element $\pi(f) = a(f) + a(f)^* \in \mathcal{A}$ is self-adjoint with respect to the involution $*$. Elements of the form $\pi(f)$ then fulfil

$$\begin{aligned}
[\pi(f), \pi(g)] &= [a(f), a(g)^*] + [a(f)^*, a(g)] \\
&= (\langle f, g \rangle - \langle g, f \rangle) \cdot I \\
&= 2i \cdot Im \langle f, g \rangle \cdot I
\end{aligned}$$

Now choose $f, g \in \mathcal{H}$ such that $Im \langle f, g \rangle = \frac{c}{2} \neq 0$, then $[\pi(f), \pi(g)] = c \cdot I$. This is possible since for any $f \in \mathcal{H} \backslash \{0\}$, we may choose $g = if$ whereby $Im \langle f, g \rangle = \|f\|^2$ whence $c = 2 \cdot \|f\|^2 \neq 0$. For an arbitrary

$k \in \mathbf{N}$ we therefore have

$$[\pi(f), \pi(g)^k] = k \cdot c \cdot \pi(g)^{k-1}$$

by successive use of (3.1). Hence

$$
\begin{aligned}
k \cdot |c| \cdot \|\pi(g)\|^{k-1} &= k \cdot |c| \cdot \|\pi(g)^{k-1}\| \\
&= \|[\pi(f), \pi(g)^k]\| \\
&\leq 2 \cdot \|\pi(f)\| \cdot \|\pi(g)^k\| \\
&= 2 \cdot \|\pi(f)\| \cdot \|\pi(g)\|^k
\end{aligned}
$$

where the C^*-norm has been properly used repeatedly together with the self-adjointness of $\pi(g)$. Hence

$$\frac{1}{2} \cdot |c| \cdot k \leq \|\pi(f)\| \cdot \|\pi(g)\|$$

for all $k \in \mathbf{N}$, so at least one of $\pi(f)$ and $\pi(g)$ is unbounded, contradicting the C^*-norm assumption. This means that we cannot turn $\mathcal{A}$ into a C^*-algebra.

It turns out (for details, see [Ya]) that there exists a normalized positive linear functional φ on $\mathcal{A}$, so we may apply the GNS construction to get a Hilbert space $\mathcal{F}$, a $*$-representation π of $\mathcal{A}$, i.e. a $*$-homomorphism from $\mathcal{A}$ into the set of linear (generally unbounded) operators on $\mathcal{F}$, and a generating vector Ω. Here $\mathcal{F}$ is the Fock space for the Fock representation π of $\mathcal{A}$ and Ω the vacuum vector. By construction $\varphi(A) = \langle \Omega, \pi(A)\Omega \rangle$, for each $A \in \mathcal{A}$. The GNS construction depends on φ. We will, however, not follow this approach but, rather, construct concrete realizations of the creation and annihilation operators on the symmetric Fock Hilbert space, and then show that it is in fact a representation of the $*$-algebra generated by $\{a(f) : f \in \mathcal{H}\}$. We thus pursue the idea used in the construction of the Fock representation of the CAR-algebra (see section 2.2 of Chap. 2).

For each $f \in \mathcal{H}$ we define the linear operator $a^*(f)$ in the symmetric Fock Hilbert space $\mathcal{F}_\mathrm{v}(\mathcal{H})$, as defined in section 2.1, by

$$a^*(f)\Omega = f \tag{3.2}$$

and on product vectors by

$$a^*(f)(f_1 \vee \cdots \vee f_n) = f \vee f_1 \vee \cdots \vee f_n \tag{3.3}$$

We ought to write $a_0^*(f)$ instead of just $a^*(f)$ as it is a specific realization (*-representation) of the abstract *-algebra $\mathcal{A}$ (as shown
below). We will, however, mainly consider this realization, so we discard the 0-indices, as we did in the case of the Fock representation
of the CAR-algebra. Extension by linearity gives a densely defined
linear operator $a^*(f)$ from $\vee^n \mathcal{H}$ to $\vee^{n+1} \mathcal{H}$, for each $n \in \mathbf{N} \cup \{0\}$.
Analogical to the CAR-case, we call $a^*(f)$ the *creation operators*, for
obvious reasons. The lack of a Pauli principle which is a consequ-
ence of the symmetry of product vectors in $\mathcal{F}_\vee(\mathcal{H})$, in contrast to the
anti-symmetry of $\mathcal{F}_\wedge(\mathcal{H})$, allows us to have non-zero product vectors
in $\mathcal{F}_\vee(\mathcal{H})$ of the form $F_n = f \vee \cdots \vee f \in \vee^n \mathcal{H}$, the n-fold (symmetric)
tensor product of $f \in \mathcal{H}$ with itself. From the abstract consideration
above we expect the creation operators defined by (3.2) and (3.3) to
become unbounded;

$$\|a^*(f)F_n\|^2 \;=\; \|\overbrace{f \vee f \vee \cdots \vee f}^{(n+1)-times}\|^2$$

$$=\; (n+1)\cdot\|f\|^2 \cdot \|\overbrace{f \vee f \vee \cdots \vee f}^{n-times}\|^2$$

$$=\; (n+1)\cdot\|f\|^2 \cdot \|F_n\|^2$$

where we have used the formula of the inner product on $\mathcal{F}_\vee(\mathcal{H})$
(see section 2.1) and that $F_n = \overbrace{f \vee \cdots \vee f}^{n-times} \in \vee^n(\mathcal{H})$. This evidently
means that $a^*(f)$ is unbounded when extended by linearity to $\mathcal{F}_\vee(\mathcal{H})$.
We will prove that in general

$$\|a^*(f)F_n\| \leq \sqrt{n+1}\cdot\|f\|\cdot\|F_n\| \tag{3.4}$$

for all $f \in \mathcal{H}$ and $F_n \in \vee^n \mathcal{H}$, where we already know that equality
occurs for a certain choice of $F_n \in \vee^n \mathcal{H}$. For $f = 0$ then $a^*(f)F_n = 0$
and the claim follows trivially. For $f \neq 0$, we put $e_1 = \frac{f}{\|f\|}$ and cho-
ose an orthonormal basis $\{e_j\}_{j=2}^\infty$ for $\{e_1\}^\perp$ in $\mathcal{H}$. Then the basis
product vectors $E_{(i)m}^{(k)m} = e_{i_1}^{k_1} \vee \cdots \vee e_{i_m}^{k_m}$, where $i_1 < \cdots < i_m$ and
$\sum_{j=1}^m k_j = n$, $k_j \in \mathbf{N}$, span a dense subspace in $\vee^n \mathcal{H}$ (see section
2.1). If we thus can prove the claim for all $F_n \in \vee^n \mathcal{H}$ of the form
$\sum_{(i)m,(k)m} a_{(i)m}^{(k)m} E_{(i)m}^{(k)m}$ then the claim holds for all $F_n \in \vee^n \mathcal{H}$. A stra-
ight forward calculation using that $\|e_{i_1}^{k_1} \vee \cdots \vee e_{i_m}^{k_m}\|^2 = \prod_{j=1}^m (k_j!)$
and that $\left\langle e_{i_1}^{k_1} \vee \cdots \vee e_{i_m}^{k_m}, e_{j_1}^{l_1} \vee \cdots \vee e_{j_p}^{l_p} \right\rangle = 0$ if $e_{i_1}^{k_1} \vee \cdots \vee e_{i_m}^{k_m} \neq$

$e_{j_1}^{l_1} \vee \cdots \vee e_{j_p}^{l_p}$, where $k_1 + \cdots + k_m = l_1 + \cdots + l_p = n$ (see section 2.1 for further details) gives

$$\left\| a^*(f) \sum a_{(i)m}^{(k)m} \cdot e_{i_1}^{k_1} \vee \cdots \vee e_{i_m}^{k_m} \right\|^2$$

$$= \|f\|^2 \cdot \left\| \sum a_{(i)m}^{(k)m} \cdot e_1 \vee e_{i_1}^{k_1} \vee \cdots \vee e_{i_m}^{k_m} \right\|^2$$

$$= \|f\|^2 \cdot \sum \left| a_{(i)m}^{(k)m} \right|^2 \cdot \left\{ \begin{array}{ll} 1 \cdot (k_1!) \cdot \ldots \cdot (k_m!) & , \text{ for } 1 < i_1 \\ (k_1 + 1)! \cdot (k_2!) \cdot \ldots \cdot (k_m!) & , \text{ for } 1 = i_1 \end{array} \right.$$

$$\leq \|f\|^2 \cdot \sum \left| a_{(i)m}^{(k)m} \right|^2 \cdot (k_1 + 1) \cdot \prod_{j=1}^{m} (k_j!)$$

$$= \|f\|^2 \cdot (k_1 + 1) \cdot \left\| \sum a_{(i)m}^{(k)m} \cdot e_{i_1}^{k_1} \vee \cdots \vee e_{i_m}^{k_m} \right\|^2$$

$$\leq (n+1) \cdot \|f\|^2 \cdot \|F_n\|^2$$

proving our claim. Then $a^*(f)$, $f \in \mathcal{H}$, are unbounded operators in $\mathcal{F}_V(\mathcal{H})$ mapping $\vee^n \mathcal{H}$ into $\vee^{n+1}\mathcal{H}$. However, $a^*(f)$ is bounded as an operator from $\vee^n \mathcal{H}$ onto $\vee^{n+1}\mathcal{H}$. Observe that $a^*(f)$ is well-defined on $\mathcal{D}(N^{1/2}) = \{F = \oplus_{n=0}^\infty F_n : \sum_{n=0}^\infty n \cdot \|F_n\|^2 < \infty\}$, where N denotes the so-called number operator given by $N(\oplus_{n=0}^\infty F_n) = \oplus_{n=0}^\infty n \cdot F_n$. In fact, $\mathcal{D}(N^{1/2})$ is the maximal domain for $a^*(f)$ on $\mathcal{F}_V(\mathcal{H})$. But in analogy with the anti-symmetric case, we may choose $\mathcal{D} = \oplus_{alg}(\vee^n \mathcal{H})$, the algebraic direct sum, as the dense domain of all $a^*(f)$, $f \in \mathcal{H}$. Indeed, $a^*(f)$ is well-defined on $\mathcal{D}$, since given any $F = \oplus_{n=0}^\infty F_n$ in $\mathcal{D}$ there exists an $N \in \mathbf{N}$ such that $F_n = 0$ for $n > N$, so $\|a^*(f)F\|^2 \leq \sum_{n=0}^N (n+1) \cdot \|f\|^2 \cdot \|F_n\|^2 \leq (N+1) \cdot \|f\|^2 \cdot \|F\|^2$. Moreover, $\mathcal{D}$ is invariant under the action of $a^*(f)$, since $a^*(f) : \vee^n \mathcal{H} \mapsto \vee^{n+1}\mathcal{H}$ creates one particle, so if $F_n = 0$, for $n > N$ then $G_n = a^*(f)F_n = 0$, for $n > N+1$, where $G = a^*(f)F = \oplus_{n=0}^\infty G_n$. In fact, $\mathcal{D}$ is a dense set of analytic vectors for all $a^*(f)$, $f \in \mathcal{H}$. For arbitrary $F_n \in \vee^n \mathcal{H}$ and $f \in \mathcal{H}$ we have

$$\|(a^*(f))^k F_n\| \leq \sqrt{\frac{(n+k)!}{n!}} \cdot \|f\|^k \cdot \|F_n\|$$

by successive use of the above norm estimate and since $a^*(f)^m F_n \in \vee^{n+m}\mathcal{H}$. Thus, for any $F = \oplus_{n=0}^\infty F_n \in \mathcal{D}$ there exists a $N \in \mathbf{N}$ such that $F_n = 0$ for $n > N$. Hence

$$\|(a^*(f))^k F\| \leq \sum_{n=0}^{N} \sqrt{\frac{(n+k)!}{n!}} \cdot \|f\|^k \cdot \|F_n\| \leq \sqrt{\frac{(N+k)!}{N!}} \cdot \|f\|^k \cdot \|F\|$$

Therefore

$$\sum_{k=0}^{\infty} \frac{\|(a^*(f))^k F\|}{k!} \cdot t^k \leq \|F\| \cdot \sum_{k=0}^{\infty} \frac{(t \cdot \|f\|)^k}{k!} \cdot \sqrt{\frac{(N+k)!}{N!}}$$

is finite for all $t \in \mathbf{R}_+$ by the ratio test. Since $\mathcal{D}$ is an invariant set of C^∞-vectors for $a^*(f)$, it follows that $\mathcal{D}$ is a set of analytic vectors for $a^*(f)$.

Now we are ready to define *annihilation operators* $a(f)$, $f \in \mathcal{H}$, on $\mathcal{D}$ (one could choose to define $a(f)$ on $\mathcal{D}(N^{1/2})$). We define annihilation operators $a(f)$, $f \in \mathcal{H}$, as the adjoint operators of $a^*(f)$, i.e. $a(f) = a^*(f)^*$, for $f \in \mathcal{H}$. Since $a^*(f)$ is unbounded, so is $a(f)$, $f \in \mathcal{H}$, and we have to specify a domain of $a(f)$. We will prove that we may choose $\mathcal{D}$ as the domain for all $a(f)$, $f \in \mathcal{H}$, and that $a(f)$ is given on product vectors in this domain by

$$a(f)\Omega = 0 \tag{3.5}$$

and

$$a(f)(f_1 \vee \cdots \vee f_n) = \sum_{i=1}^{n} \langle f, f_i \rangle \cdot f_1 \vee \cdots \vee f_{i-1} \vee f_{i+1} \vee \cdots \vee f_n \tag{3.6}$$

Observe that $a(f)$ is bounded as an operator from $\vee^{n+1}\mathcal{H}$ to $\vee^n\mathcal{H}$, since $a^*(f)$ is a bounded operator from $\vee^n\mathcal{H}$ to $\vee^{n+1}\mathcal{H}$. Since $\langle \Omega, a^*(f)F \rangle = 0$ for all $F \in \mathcal{D}$, it follows trivially that $\Omega \in \mathcal{D}(a(f))$ and that $a(f)\Omega = 0$. Let $f = f_1 \in \mathcal{H}$ be arbitrarily chosen. For

$$F_n = f_2 \vee \cdots \vee f_{n+1} \in \vee^n\mathcal{H}, \quad n \in \mathbf{N}$$

and

$$G_{n+1} = g_1 \vee \cdots \vee g_{n+1} \in \vee^{n+1}\mathcal{H}$$

it follows

$$\langle G_{n+1}, a^*(f_1)F_n \rangle$$
$$= \langle g_1 \vee \cdots \vee g_{n+1}, f_1 \vee f_2 \cdots \vee f_{n+1} \rangle$$
$$= \sum_{i=1}^{n+1} \langle g_i, f \rangle \cdot \langle G_n^{(i)}, F_n \rangle$$
$$= \left\langle \sum_{i=1}^{n+1} \langle g_i, f \rangle \cdot g_1 \vee \cdots \vee g_{i-1} \cdot g_{i+1} \vee \cdots \vee g_{n+1}, F_n \right\rangle$$

where $G_n^{(i)}$ denotes $g_1 \vee \cdots \vee g_{i-1} \vee g_{i+1} \vee \cdots \vee g_{n+1}$ and we have used the formula for the inner product on $\mathcal{F}_V(\mathcal{H})$ (see section 2.1). Hence formula (3.6) follows. We have thus shown that the operator defined by (3.5) and (3.6) fulfils that $a(f) \subset a^*(f)^*$. We may therefore choose $\mathcal{D}$ as the dense domain of all the $a(f)$, $f \in \mathcal{H}$. Note that the mapping $f \mapsto a(f)$ evidently is anti-linear. Since $a^*(f)$ is bounded as an operator from $\vee^n \mathcal{H}$ onto $\vee^{n+1} \mathcal{H}$, so is $a(f) = a^*(f)^*$ as an operator from $\vee^{n+1} \mathcal{H}$ onto $\vee^n \mathcal{H}$, and we have

$$
\begin{aligned}
\|a(f)\|_{\vee^{n+1}\mathcal{H}}\|^2 &= \|a^*(f)\|_{\vee^n\mathcal{H}}\|^2 \\
&= (n+1) \cdot \|f\|^2
\end{aligned}
$$

by (3.4). Therefore

$$
\|a(f)F_n\| \leq \sqrt{n} \cdot \|f\| \cdot \|F_n\| \tag{3.7}
$$

for any $F_n \in \vee^n \mathcal{H}$ and $f \in \mathcal{H}$. Observe that the norm estimate for $a(f)$ is almost the same as the norm estimate for $a^*(f)$, except for a factor $\sqrt{\frac{n}{n+1}}$. We already know that the domain $\mathcal{D}$ for $a(f)$ is dense in $\mathcal{F}_V(\mathcal{H})$ and since $a(f) : \vee^n \mathcal{H} \mapsto \vee^{n-1} \mathcal{H}$ it follows that $\mathcal{D}$ is invariant. In fact, $\mathcal{D}$ is a set of analytic vectors for all $a(f)$, $f \in \mathcal{H}$, since for any $F = \oplus_{n=0}^\infty F_n \in \mathcal{D}$ there exists a $N \in \mathbf{N}$ such that $F_n = 0$ for $n > N$. Therefore $(a(f))^k F = 0$, for $k > N$. Hence

$$
\sum_{k=0}^\infty \frac{\|(a(f))^k F\|}{k!} \cdot t^k = \sum_{k=0}^N \frac{\|(a(f))^k F\|}{k!} \cdot t^k
$$

which evidently is finite for all $t \in \mathbf{R}_+$. We call $a(f)$, $f \in \mathcal{H}$, *annihilation operators*, for obvious reasons. As in the case of the creation operator, we could define $a(f)$ on $\mathcal{D}(N^{1/2})$ which is the maximal domain of $a(f)$ on $\mathcal{F}_V(\mathcal{H})$.

We will now verify that $a^*(f)$ and $a(g)$, $f, g \in \mathcal{H}$ fulfil the canonical commutation relations on $\mathcal{D}$. In fact we merely have to prove the claim on product vectors and the vacuum vector, since a finite linear combination of these spans a dense set of $\mathcal{F}_V(\mathcal{H})$. Let $n \in \mathbf{N}$ be arbitrarily chosen and consider any $F_n \in \vee^n \mathcal{H}$ of the form $F_n = f_1 \vee \cdots \vee f_n$. Then

$$
\begin{aligned}
a(g)a^*(f)(f_1 \vee \cdots \vee f_n) &= a(g)(f \vee f_1 \vee \cdots \vee f_n) \\
&= \langle g, f \rangle F_n + \sum_{i=1}^n \langle g, f_i \rangle G_n^{(i)}
\end{aligned}
$$

where $G_n^{(i)} = f \vee f_1 \vee \cdots \vee f_{i-1} \vee f_{i+1} \vee \cdots \vee f_n$, for $i = 1, \ldots, n$. On the other hand

$$a^*(f)a(g)(f_1 \vee \cdots \vee f_n) = a^*(f)\left(\sum_{i=1}^{n} \langle g, f_i \rangle F_n^{(i)}\right)$$

$$= \sum_{i=1}^{n} \langle g, f_i \rangle G_n^{(i)}$$

where $F_n^{(i)} = f_1 \vee \cdots \vee f_{i-1} \vee f_{i+1} \vee \cdots \vee f_n$, for $i = 1, \ldots, n$ and $G_n^{(i)}$ is as above. It immediately follows that

$$[a(g), a^*(f)] = \langle g, f \rangle \cdot I$$

on $\vee^n \mathcal{H}$. Moreover, we have $a(g)a^*(f)\Omega = \langle g, f \rangle \Omega$ and $a^*(f)a(g)\Omega = a^*(f)0 = 0$. Hence it follows that

$$[a(g), a^*(f)] = \langle g, f \rangle \cdot I$$

on $\mathcal{D}$. Finally, we have

$$[a^*(g), a^*(f)]f_1 \vee \cdots \vee f_n = g \vee f \vee f_1 \vee \cdots \vee f_n - f \vee g \vee f_1 \vee \cdots \vee f_n = 0$$

by the symmetry, hence also $[a(g), a(f)] = 0$ on $\mathcal{D}$. Trivially, $[a(g), a(f)]\Omega = 0$ and $[a^*(g), a^*(f)]\Omega = 0$ as well. Thus

$$[a^*(g), a^*(f)] = 0 = [a(g), a(f)]$$

on $\mathcal{D}$. Thus, the commutation relations also hold on the maximal domain $\mathcal{D}(N^{1/2})$. We summarize this in the following theorem.

Theorem 9 *The unbounded creation and annihilation operators on the symmetric Fock Hilbert space $\mathcal{F}_\vee(\mathcal{H})$ with domain $\mathcal{D}$ or $\mathcal{D}(N^{1/2})$ as constructed above have $\mathcal{D}$ as a dense set of analytical vectors and fulfil the canonical commutation relations and thereby give a *-representation of the *-algebra generated by $\{a(f) : f \in \mathcal{H}\}$.*

Proof. A direct consequence of the discussion above. However, some of these statements are also discussed and proved in [B-R 2, section 5.2.1]. □

We call this *-representation for the Fock representation of the *-algebra generated by $\{a(f) : f \in \mathcal{H}\}$. Observe that the vacuum vector Ω is a cyclic vector, since each product vector $f_1 \vee \cdots \vee f_n$ is of the form $a^*(f_1) \cdot \ldots \cdot a^*(f_n)\Omega$.

Define

$$\pi(f) = \frac{1}{\sqrt{2}}(a(f) + a^*(f)) \tag{3.8}$$

for all $f \in \mathcal{H}$, with domain $\mathcal{D}$. Following [R-S 1, p.209] the operator $\pi(f)$ is called *the Segal field operator* and the mapping $f \to \pi(f)$ is called *the Segal quantization over* $\mathcal{H}$. Hence $a(f)$ and $a^*(f)$ can be recovered from $\pi(f)$ by the formula $a^*(f) = \frac{1}{\sqrt{2}}(\pi(f) - i\pi(if))$ and $a(f) = \frac{1}{\sqrt{2}}(\pi(f) + i\pi(if))$. Accordingly, there is a one-to-one correspondence between the two points of view, and we may therefore consider the $*$-algebra generated by $\{\pi(f) : f \in \mathcal{H}\}$. Notice that the mapping $f \mapsto \pi(f)$ is only real linear. From above it immediately follows that Ω is a cyclic vector, i.e. the linear span of

$$\{\pi(f_1) \cdot \ldots \cdot \pi(f_n)\Omega : f_1, \ldots, f_n \in \mathcal{H}, \ n \in \mathbf{N}\}$$

is dense in $\mathcal{F}_\vee(\mathcal{H})$. Moreover, the canonical commutation relations for $a(f)$ and $a(g)$, $f, g \in \mathcal{H}$, give

$$\begin{aligned} [\pi(f), \pi(g)] \\ = \ & i \cdot Im \langle f, g \rangle \cdot I \\ = \ & i \cdot \sigma(f, g) \cdot I \end{aligned} \tag{3.9}$$

on $\mathcal{D}$, where $\sigma(f, g) = Im \langle f, g \rangle$ is a non-degenerated symplectic bilinear form on $\mathcal{H}$ as a real Hilbert space. The restriction of $\pi(f)$ to $\vee^n \mathcal{H}$ is bounded, since both $a(f)$ and $a^*(f)$ are bounded. In fact

$$\begin{aligned} \|\pi(f)F_n\| \ & \leq \ \frac{1}{\sqrt{2}} (\|a(f)F_n\| + \|a^*(f)F_n\|) \\ & \leq \ \sqrt{2 \cdot (n+1)} \|f\| \cdot \|F_n\| \end{aligned} \tag{3.10}$$

Then by induction after $k \in \mathbf{N}$ we get

$$\|\pi(f)^k F_n\| \leq 2^{\frac{k}{2}} \sqrt{\frac{(n+k)!}{n!}} \cdot \|f\|^k \cdot \|F_n\|$$

where we will use that $\pi(f) : \vee^n \mathcal{H} \to \vee^{n-1}\mathcal{H} \oplus \vee^{n+1}\mathcal{H}$ and the equations (3.4) and (3.7). Assume that $\|\pi(f)^k F_n\| \leq 2^{\frac{k}{2}} \cdot \sqrt{\frac{(n+k)!}{n!}}$.

$\|f\|^k \cdot \|F_n\|$ then

$$\|\pi(f)^{k+1} F_n\| \leq \frac{1}{\sqrt{2}} \left(\|\pi(f)^k (a(f)F_n)\| + \|\pi(f)^k (a^*(f)F_n)\| \right)$$

$$\leq 2^{\frac{k+1}{2}} \cdot \sqrt{\frac{(n+(k+1))!}{n!}} \cdot \|f\|^{k+1} \cdot \|F_n\|$$

proving the formula for $k+1$. Notice that the formula is valid for $k = 0$ too and is proven above for $k = 1$.

Therefore

$$\sum_{k=0}^{\infty} \frac{t^k}{k!} \|\pi(f)^k F_n\| \leq \sum_{k=0}^{\infty} \frac{(\sqrt{2} \cdot t)^k}{k!} \sqrt{\frac{(n+k)!}{n!}} \cdot \|f\|^k \cdot \|F_n\|$$

is finite for all $t \in \mathbf{R}_+$, by the ratio test. Then each $F_n \in \vee^n \mathcal{H}$ and each $\oplus_{n=0}^{\infty} F_n \in \mathcal{D}$ is an analytic vector for $\pi(f)$. Hence $\pi(f)$ is essentially self-adjoint on $\mathcal{D}$, for each $f \in \mathcal{H}$, by Nelsons analytic vector theorem (see [R-S 2; p. 202]), since $\pi(f)$ is indeed a symmetric operator. Notice that the mapping $f \to \pi(f)$ is strongly continuous, by

$$\|(\pi(f) - \pi(g))F\| = \|\pi(f-g)F\| \leq \sqrt{2} \cdot \|f - g\| \cdot \left\| (N+1)^{\frac{1}{2}} F \right\|$$

for $F \in \mathcal{D}$, where N denotes the number operator mentioned earlier.

The $*$-algebra generated by $\{\pi(f) : f \in \mathcal{H}\}$ is usually denoted the Fock representation of the corresponding abstract $*$-algebra.

We will now use the self-adjoint operators $\overline{\pi(f)}$, the closure of the essentially self-adjoint operators $\pi(f)$, to construct the $Weyl$ $operators$ $W(f)$, $f \in \mathcal{H}$, as the unitary operators on $\mathcal{F}_\vee(\mathcal{H})$ given by

$$W(f) = e^{i\overline{\pi(f)}} \qquad (3.11)$$

for $f \in \mathcal{H}$. Henceforth we will just write $\pi(f)$ for the self-adjoint closure $\overline{\pi(f)}$ of $\pi(f)$. Therefore $\pi(f)$ denotes two different operators interchangeably, one operator being the closed extension of the other. If it is not evident from the context to which operator $\pi(f)$ is referring, we explicitly mention it.

We then get the following proposition, due to E. Segal.

Proposition 10 *Let $f, g \in \mathcal{H}$. Then the Weyl operators $W(f)$ and $W(g)$ fulfil*

1) $W(f)W(g) = W(f+g)e^{-\frac{1}{2}\cdot i \cdot \sigma(f,g)}$

2) $W(f)$ is strongly continuous in f

3) $\langle \Omega, W(f)\Omega \rangle = e^{-\frac{1}{4}\cdot \|f\|^2}$

Proof. This proposition can also be found in [B-R 2, theorem 5.2.4 (p.13) and on p.25]. However, our proof is along different lines.

1) From equation (3.9) it follows by induction for arbitrary $f, g \in \mathcal{H}$ that

$$\frac{1}{n!}\left(\pi(f) + \pi(g)\right)^n =$$

$$\sum_{k+l+2\cdot m=n} \frac{1}{k!}\cdot\pi(f)^k \cdot \frac{1}{l!}\cdot\pi(g)^l\frac{1}{m!}\cdot\left(-\frac{1}{2}\cdot i \cdot \sigma(f,g)\right)^m$$

on $\mathcal{D}$, for any $n \in \mathbf{N}\cup\{0\}$. For $n = 0$, both sides yield I. For $n = 1$, the right-hand side gives $\pi(f)+\pi(g)$, which obviously equals the left hand side. Assume that the formula holds for any $n = N \in \mathbf{N}$. This implies that

$$\frac{1}{N!}\left(\pi(f) + \pi(g)\right)^{N+1}$$

$$= \left(\pi(f) + \pi(g)\right) \cdot \frac{1}{N!}\cdot \left(\pi(f) + \pi(g)\right)^N$$

$$= \left(\pi(f) + \pi(g)\right)\sum_{k+l+2\cdot m=N} \frac{1}{k!}\pi(f)^k \cdot \frac{1}{l!}\pi(g)^l \cdot \frac{1}{m!}\left(-\frac{1}{2}i\sigma(f,g)\right)^m$$

$$= \sum_{k+l+2\cdot m=N} \left(\frac{1}{k!}\cdot\pi(f)^{k+1} \cdot \frac{1}{l!}\cdot\pi(g)^l \cdot \frac{1}{m!}\cdot\left(-\frac{1}{2}\cdot i \cdot \sigma(f,g)\right)^m\right.$$

$$+\frac{1}{k!}\cdot\pi(f)^k \cdot \frac{1}{l!}\cdot\pi(g)^{l+1} \cdot \frac{1}{m!}\cdot\left(-\frac{1}{2}\cdot i \cdot \sigma(f,g)\right)^m$$

$$\left.+ k\cdot\frac{1}{k!}\cdot\pi(f)^{k-1} \cdot \frac{1}{l!}\cdot\pi(g)^l \cdot \frac{1}{m!}\cdot 2\left(-\frac{1}{2}\cdot i \cdot \sigma(f,g)\right)^{m+1}\right)$$

where we have used equation (3.9) together with $\sigma(g, f) = -\sigma(f,g)$, whereby

$$\left[\pi(g), \pi(f)^k\right] = k \cdot 2\left(-\frac{1}{2}\cdot i \cdot \sigma(f,g)\right)\pi(f)^{k-1}$$

on $\mathcal{D}$, for every $k \in \mathbf{N}$. Hence we get

$$\frac{1}{N!} \left(\pi(f) + \pi(g) \right)^{N+1}$$

$$= \sum_{k+l+2\cdot m=N+1} \left(k \cdot \frac{1}{k!} \cdot \pi(f)^k \cdot \frac{1}{l!} \cdot \pi(g)^l \cdot \frac{1}{m!} \cdot \left(-\frac{1}{2} \cdot i \cdot \sigma(f,g) \right)^m \right.$$

$$+ l \cdot \frac{1}{k!} \cdot \pi(f)^k \cdot \frac{1}{l!} \cdot \pi(g)^l \cdot \frac{1}{m!} \cdot \left(-\frac{1}{2} \cdot i \cdot \sigma(f,g) \right)^m$$

$$\left. + 2m \cdot \frac{1}{k!} \cdot \pi(f)^k \cdot \frac{1}{l!} \cdot \pi(g)^l \cdot \frac{1}{m!} \cdot \left(-\frac{1}{2} \cdot i \cdot \sigma(f,g) \right)^m \right)$$

$$= (N+1) \sum_{k+l+2\cdot m=N+1} \frac{1}{k!} \pi(f)^k \cdot \frac{1}{l!} \pi(g)^l \cdot \frac{1}{m!} \left(-\frac{1}{2} \cdot i \cdot \sigma(f,g) \right)^m$$

from which the desired formula immediately follows for $n = N + 1$, by dividing by $N+1$, whereby the induction is completed. It follows by Cauchy-Schwarz's inequality that the double series

$$\sum_{k,l=0}^{\infty} \frac{1}{k!} \cdot \frac{1}{l!} \langle \pi(g)^l G, \pi(f)^k F \rangle$$

converges absolutely for any $F, G \in \mathcal{D}$. Now the result given above implies that

$$\langle W(f+g)G, F \rangle$$

$$= \left\langle \sum_{n=0}^{\infty} \frac{i^n}{n!} \pi(f+g)^n G, F \right\rangle$$

$$= \sum_{n=0}^{\infty} \left\langle \frac{i^n}{n!} \left(\pi(f) + \pi(g) \right)^n G, F \right\rangle$$

$$= \sum_{n=0}^{\infty} \sum_{k+l+2\cdot m=n} \left\langle \frac{i^l}{l!} \pi(g)^l G, \frac{i^k}{k!} \pi(f)^k F \right\rangle \cdot \frac{1}{m!} \left(-\frac{1}{2} i \sigma(f,g) \right)^m$$

$$= \left\langle \sum_{l=0}^{\infty} \frac{i^l}{l!} \pi(g)^l G, \sum_{k=0}^{\infty} \frac{i^k}{k!} \pi(f)^k F \right\rangle \sum_{m=0}^{\infty} \frac{1}{m!} \left(-\frac{1}{2} i \sigma(f,g) \right)^m$$

$$= \langle W(g)G, W(f)F \rangle \cdot e^{-\frac{1}{2} \cdot i \cdot \sigma(f,g)}$$

$$= \left\langle W(f) W(g) \cdot e^{-\frac{1}{2} \cdot i \cdot \sigma(f,g)} G, F \right\rangle$$

for all $F, G \in \mathcal{D}$. Since $\mathcal{D}$ is dense in $\mathcal{H}$ it follows that

$$W(f) W(g) = W(f+g) \cdot e^{-\frac{1}{2} \cdot i \cdot \sigma(f,g)}$$

for any $f, g \in \mathcal{H}$, proving 1). Notice that we have obtained

$$W(f)W(g) = W(g)W(f) \cdot e^{-i \cdot \sigma(f,g)}$$

on $\mathcal{H}$ and that $W(-f)W(f) = W(0) = I$, for all $f, g \in \mathcal{H}$, as a corrolary.

2) For any $F \in \mathcal{D}$ there exists an $n \in \mathbf{N}$ such that $F_n = 0$, for $n > N$, where $F = \oplus_{n=0}^{\infty} F_n$. The previously derived norm estimate

$$\left\| \pi(f)^k F_n \right\| \leq 2^{\frac{k}{2}} \cdot \sqrt{\frac{(n+k)!}{n!}} \cdot \|f\|^k \cdot \|F_n\|$$

then gives $\left\| \pi(f)^k F_n \right\| \leq 2^{\frac{k}{2}} \cdot \sqrt{\frac{(N+k)!}{N!}} \cdot \|f\|^k \cdot \|F\|$, for any $f \in \mathcal{H}$. For any $f \in \mathcal{H}$ and any $F \in \mathcal{D}$ we therefore have

$$
\begin{aligned}
\|(W(f) - 1)F\| &= \left\| \sum_{k=1}^{\infty} \frac{1}{k!} \pi(f)^k F \right\| \\
&\leq \sum_{k=1}^{\infty} \frac{1}{k!} \cdot 2^{\frac{k}{2}} \cdot \sqrt{\frac{(N+k)!}{N!}} \cdot \|f\|^k \cdot \|F\|
\end{aligned}
$$

which converges, by the ratio test, with a limit less than the convergent series $\sum_{k=1}^{\infty} \frac{1}{k!} \cdot 2^{\frac{k}{2}} \cdot \sqrt{\frac{(N+k)!}{N!}}$ times $\|f\| \cdot \|F\|$, for f in the unit ball. Hence $\|(W(f) - 1)F\| \to 0$, as $f \to 0$, i.e. $W(f)$ is strongly continuous at $f = 0$ on $\mathcal{D}$, and then on all of $\mathcal{H}$.

For arbitrary $f, g \in \mathcal{H}$, we obtain the following

$$
\begin{aligned}
\|(W(f) &- W(g))F\| \\
&= \|W(g)(W(-g)W(f) - 1)F\| \\
&\leq \left\| \left(W(f-g) \cdot e^{-\frac{1}{2} \cdot i \cdot \sigma(f,g)} - 1 \right) F \right\| \\
&\leq \|(W(f-g) - 1) F\| + \left| 1 - e^{+\frac{1}{2} \cdot i \cdot \sigma(f,g)} \right| \cdot \|F\| \\
&\to 0 \quad \text{as } g \to f
\end{aligned}
$$

by use of **1)**, the strong continuity of $W(f)$ at zero, the continuity of $\sigma(f,g) = Im \langle f, g \rangle$ and the fact that $\sigma(f,f) = 0$. Hence $W(f)$ is strongly continuous in $f \in \mathcal{H}$, proving **2)**.

3) We first notice that $[a(f), \pi(f)] = \frac{1}{\sqrt{2}} \|f\|^2 \cdot I$, by equation (3.1) and (3.8). This implies

$$[a(f), \pi(f)^k] = k \cdot \frac{1}{\sqrt{2}} \cdot \|f\|^2 \cdot \pi(f)^{k-1} \tag{3.12}$$

for all $k \in \mathbf{N}$, on $\mathcal{D}$. For $n \geq 2$ we get

$$
\begin{aligned}
\langle \Omega, \pi(f)^n \Omega \rangle &= \langle \pi(f)\Omega, \pi(f)^{n-1}\Omega \rangle \\
&= \frac{1}{\sqrt{2}} \langle a^*(f)\Omega, \pi(f)^{n-1}\Omega \rangle \\
&= \frac{1}{\sqrt{2}} \langle \Omega, a(f)\pi(f)^{n-1}\Omega \rangle \\
&= \frac{1}{\sqrt{2}} \left\langle \Omega, (n-1)\frac{1}{\sqrt{2}} \|f\|^2 \cdot \pi(f)^{n-2}\Omega \right\rangle \\
&= \frac{n-1}{2} \|f\|^2 \langle \Omega, \pi(f)^{n-2}\Omega \rangle \qquad\qquad (3.13)
\end{aligned}
$$

where we have used equation (3.12) for $k = n - 1$.

Observe now that

$$
\langle \Omega, \pi(f)^0 \Omega \rangle = \langle \Omega, \Omega \rangle = 1
$$

$$
\langle \Omega, \pi(f)\Omega \rangle = \left\langle \Omega, \frac{1}{\sqrt{2}}f \right\rangle = 0
$$

$$
\langle \Omega, \pi(f)^2 \Omega \rangle = \|\pi(f)\Omega\|^2 = \frac{1}{2}\|f\|^2
$$

and

$$
\langle \Omega, \pi(f)^3 \Omega \rangle = \langle f, \|f\|^2 \Omega + f \vee f \rangle = 0
$$

Then, iteration by use of equation (3.13) and the recently derived properties give us that

$$
\langle \Omega, \pi(f)^n \Omega \rangle = \begin{cases} 0 & , \quad \text{for } n \text{ odd} \\ \frac{(n-1)(n-3)\cdots 1}{2^{n/2}} \cdot \|f\|^n & , \quad \text{for } n \text{ even} \end{cases}
$$

Rewriting $(2n-1)(2n-3) \cdot \ldots \cdot 1$ as $\frac{(2n)!}{2^n \cdot n!}$ one obtains

$$
\langle \Omega, \pi(f)^{2n+1}\Omega \rangle = 0
$$

and

$$
\langle \Omega, \pi(f)^{2n}\Omega \rangle = \frac{(2n)!}{2^{2n} \cdot n!} \cdot \|f\|^{2n}
$$

for $n \in \mathbf{N} \cup \{0\}$. Since Ω is an analytic vector for $\pi(f)$, we may use the following expansion

$$
\langle \Omega, W(f)\Omega \rangle = \sum_{n=0}^{\infty} \frac{i^n}{n!} \langle \Omega, \pi(f)^n \Omega \rangle
$$

$$= \sum_{n=0}^{\infty} \frac{(-1)^n}{(2n)!} \cdot \frac{(2n)!}{2^{2n} \cdot n!} \cdot \|f\|^{2n}$$

$$= \sum_{n=0}^{\infty} \frac{1}{n!} (-\frac{1}{4} \cdot \|f\|^2)^n$$

$$= e^{-\frac{1}{4} \cdot \|f\|^2}$$

by use of the equations derived above, proving **3**). We notice that claim **2**) also follows from **3**) by direct calculation. □

The specific C^*-algebra, generated by the Weyl operators $W(f)$, $f \in \mathcal{H}$, is called the *Fock representation of the CCR-algebra*. For completeness reasons, we briefly discuss the (abstract) CCR-algebra below, though it is not needed for the argumentation to come. We will, however, state some propositions first.

Proposition 11 *The Fock space ∗-representations $a(f)$ and $\pi(f)$ are irreducible, in the sense that any bounded operator T on $\mathcal{F}_{\vee}(\mathcal{H})$ commuting with the ∗-algebra generated by $a(f)$ and $\pi(f)$, respectively, is trivial, i.e. T is a scalar multiple of the unit operator.*

Proof. Since $a(f) = \frac{1}{\sqrt{2}}(\pi(f) + i\pi(if))$ and $a^*(f) = \frac{1}{\sqrt{2}}(\pi(f) - i\pi(if))$, any operator T commuting with $\pi(f)$, $f \in \mathcal{H}$, commutes with $a(f)$ and $a^*(f)$, and conversely. Therefore we only need to consider one of the statements, since they imply each other. Let T commute with $a(f)$ and $a^*(f)$ on $\mathcal{D}$, i.e. $T\mathcal{D} \subset \mathcal{D}$ and $[T, a(f)] = [T, a^*(f)] = 0$, on $\mathcal{D}$. Then the rest of the proof is quite similar to that of theorem 1 □

Corollary 12 *The Fock representation $W(f)$, $f \in \mathcal{H}$, of the CCR-algebra is irreducible.*

Proof. For a proof, see [R-S 2, p.232]. It seems that the argument of the proof in [B-R 2, p.13] fails. □

In analogy to the treatment of the CAR-algebra (see section 2.2 and 2.4), we may consider $\mathcal{H}$ as a real Hilbert space, equipped with a non-degenerate symplectic bilinear form σ, i.e. $\sigma(f, g) = -\sigma(g, f)$, for all $f, g \in \mathcal{H}$, and if $f \in \mathcal{H} \setminus \{0\}$ there exists a $g \in \mathcal{H}$ such that $\sigma(f, g) \neq 0$. Notice that the σ previously defined in (3.9), is an example of such a form. We are now ready to define *the (abstract) CCR-algebra* (following [B-R 2, p.20]).

Definition 13 *Let $\mathcal{H}$ be a real Hilbert space equipped with a non-degenerate symplectic bilinear form σ, then the CCR-algebra is the C^*-algebra generated by non-zero elements $W(f)$, $f \in \mathcal{H}$, satisfying $W(-f) = W(f)^*$ and $W(f)W(g) = e^{-\frac{1}{2} \cdot i \cdot \sigma(f,g)} \cdot W(f + g)$ for all $f, g \in \mathcal{H}$.*

Notice that the previously constructed specific C^*-algebra, generated by the Weyl operators, fulfils the conditions of being a CCR-algebra. Observe that we indicate uniqueness by use of "the" instead of "a" in "the CCR-algebra". This is justified by the following theorem.

Theorem 14 *Let $\mathcal{A}_1$ and $\mathcal{A}_2$ be two CCR-algebras, then there exists a unique $*$-isomorphism $\alpha : \mathcal{A}_1 \mapsto \mathcal{A}_2$ mapping the generators $W_1(f)$ of $\mathcal{A}_1$ into the generators $W_2(f)$ of $\mathcal{A}_2$, for all $f \in \mathcal{H}$. Thus the CCR-algebra is unique, up to $*$-isomorhism. Moreover, $W(0) = 1$ and $W(f)$ is unitary, for all $f \in \mathcal{H}$. Furthermore, if S is a real linear invertible operator on $\mathcal{H}$, leaving σ invariant, i.e. $\sigma(Sf, Sg) = \sigma(f,g)$ for all $f, g \in \mathcal{H}$, then there exists a unique $*$-automorphism α on the CCR-algebra, such that $\alpha(W(f)) = W(S^{-1}f)$, for all $f \in \mathcal{H}$.*

Proof. We will not prove this theorem except for the last part, since we won't need the rest throughout this book. For a detailed proof of the theorem consult [B-R 2, p. 20-22.]. However, the last part follows directly from the first part, with $W_1(f) = W(f)$ and $W_2(f) = W(S^{-1}f)$, $f \in \mathcal{H}$. Such real, linear, invertible operators S on $\mathcal{H}$ leaving σ invariant, are usually called Boguliubov transformations.

    The last part can be shown directly, though. We only have to show that $W_2(f) = W(S^{-1}f)$ fulfils the Weyl form of the canonical anti-commutation relations, given in 1) of proposition 10. We get $W(S^{-1}f)W(S^{-1}g) = W(S^{-1}(f + g)) \cdot e^{-\frac{1}{2}i\sigma(S^{-1}f, S^{-1}g)} = W(S^{-1}(f+g)) \cdot e^{-\frac{1}{2}i\sigma(f,g)}$ that is $W_2(f)$, $f \in \mathcal{H}$, form a CCR-algebra by definition 13, so the mapping $W(f) \mapsto W(S^{-1}f)$ defines a unique $*$-morphism. $\qquad\qquad\square$

    The generalization of the Fock representation of the CCR-algebra above is somehow very slight. Let σ be a non-degenerate symplectic form on a real Hilbert space $\mathcal{H}_r$, for example, $\mathcal{H}$ considered as a real Hilbert space. Then there exists a real linear operator J on $\mathcal{H}_r$ such that $J^2 = -I$ and $\sigma(Jf,g) = -\sigma(f, Jg)$ for all $f, g \in \mathcal{H}_r$. In fact, if

$\{u_k, v_k\}_{k\in\mathbf{N}}$ is a symplectic basis for $\mathcal{H}_r$, i.e. $\sigma(u_k, u_l) = \sigma(v_k, v_l) = 0$ and $\sigma(u_k, v_l) = \delta_{k-l}$, for all $k, l \in \mathbf{N}$, then we put $Ju_k = v_k$ and $Jv_k = -u_k$. Extension of J by (real) linearity and continuity gives us an operator with the above properties on $\mathcal{H}_r$ (one can always choose a symplectic basis for an even or infinite-dimensional space, see for example [Arn, p. 220]). Such an operator J introduce a complex structure on $\mathcal{H}_r$, and reflects the complex structure of a complex Hilbert space $\mathcal{H}$. The correspondence between $\mathcal{H}$ and $\mathcal{H}_r$ is then given by $e_k \leftrightarrow u_k$ and $i \cdot e_k \leftrightarrow v_k = Ju_k$, where $\{e_k = u_k\}_{k\in\mathbf{N}}$ is a basis of $\mathcal{H}$, and $\langle f, g \rangle = \sigma(f, Jg) + i\sigma(f, g)$ is an inner product on $\mathcal{H}$. Observe that $\sigma(f, g) = Im\,\langle f, g \rangle$, for all $f, g \in \mathcal{H}$, and $\mathcal{H}_r$ is $\mathcal{H}$ considered as a real Hilbert space.

We now turn to a construction called the second quantization in $\mathcal{F}_V(\mathcal{H})$, in the following section.

3.2   The Second Quantization in $\mathcal{F}_V(\mathcal{H})$

In this section we describe the second quantization, based on the explicit Fock Hilbert space construction in the symmetric case, given in section 2.1, and on the Fock $*$-representation of the $*$-algebra generated by $\{a(f) : f \in \mathcal{H}\}$, given in section 3.1, following the pattern outlined in the anti-symmetric case, given in section 2.3. This section will be used frequently in the construction of the so-called metaplectic representation, to be treated in the next section.

As mentioned in section 2.3, the basic idea of the second quantization goes back to V. Fock ([Fo]) in 1932. However, it was J.M Cook ([Co]) who in 1953 made the construction in detail. The construction is a method of lifting one-particle operators on a Hilbert space to many-particle operators on the Fock Hilbert space, whenever it is possible (in the symmetric case one uses the symmetric Fock Hilbert space). It gives a nice mathematical frame for handling many problems in quantum physics with infinitely many degrees of freedom.

Our treatment of second quantization is somewhat different from those given by Cook in [Co] and by Bratteli and Robinson in [B-R 2].

As in section 2.3 one defines a strongly continuous one-parameter unitary group of operators $U(A)_n$ on $\vee^n\mathcal{H}$, for any bounded skew-self-adjoint operator A, given by $U(A)_0\Omega = \Omega$ and

$$U(A)_n(f_1 \vee \ldots \vee f_n) = e^A f_1 \vee \ldots \vee e^A f_n \qquad (3.14)$$

on product vectors $f_1 \vee \ldots \vee f_n \in \vee^n \mathcal{H}$. The infinetesimal generators $dU(A)_n = \frac{d}{dt}\big|_{t=0} U(tA)_n$ on its domain fulfil

$$U(tA)_n = e^{t \cdot dU(A)_n} \tag{3.15}$$

and become

$$dU(A)_n F_n = \sum_{i=1}^{n} f_1 \vee \ldots \vee f_{i-1} \vee A f_i \vee f_{i+1} \vee \ldots \vee f_n \tag{3.16}$$

on product vectors. Moreover, $dU(A)_0 \Omega = 0$.

Similar to our preceding notational conventions, we let $\mathcal{D}$ denote the algebraic direct sum $\oplus_{alg} \vee^n \mathcal{H}$. It then follows that each $dU(A)_n$ leaves $\vee^n \mathcal{H}$ invariant and that $dU(A) = \oplus_{n=0}^{\infty} dU(A)_n$ is well-defined on $\mathcal{D}$. Observe that $dU(A)$ is skew-symmetric with a dense invariant domain $\mathcal{D}$.

Consider now possible unbounded skew-self-adjoint operators A. We define $dU(A)_n$ directly by $dU(A)_0 \Omega = 0$ and by (3.16) on product vectors in $\vee^n \mathcal{H}$, such that each one-particle vector $f_1, \ldots, f_n$ belongs to the domain $\mathcal{D}(A)$ of A. We denote this set of product vectors by $\mathcal{D}(A)^{\vee n}$, for each $n \in \mathbb{N} \cup \{0\}$. Then $dU(A)_n$ is well-defined on $\mathcal{D}(A)^{\vee n}$. Put $dU(A) = \oplus_{n=0}^{\infty} dU(A)_n$ on the algebraic direct sum $\oplus_{alg} \mathcal{D}(A)^{\vee n}$.

In Complete analogy with the anti-symmetric case one can show

$$\sum_{k=0}^{\infty} \frac{1}{k!} \left\| dU(A)^k (f_1 \vee \ldots \vee f_n) \right\|$$

$$\leq \sum_{k=0}^{\infty} \frac{(n \cdot M)^k}{k!} \cdot \|f_1 \otimes \cdots \otimes f_n\| \cdot (n!)^{\frac{1}{2}}$$

$$= (n!)^{\frac{1}{2}} \cdot e^{n \cdot M} \cdot \|f_1 \otimes \cdots \otimes f_n\|$$

$$< \infty$$

on any product vector of analytic vectors for A. This means that the set $\mathcal{D}_A$ of finite linear combinations of Ω and n-folded tensor products $f_1 \vee \ldots \vee f_n$, $n \in \mathbb{N}$, such that each one-particle vector f_j is an analytic vector for A, is a set of analytic vectors for $dU(A)$.

We observe that $\mathcal{D}_A$ is a dense set, since the analytic vectors for A in $\mathcal{D}(A)$ form a dense set in $\mathcal{H}$. Then Nelson's theorem (transformed to essentially skew-self-adjoint operators) state that the operator $dU(A)$ is essentially skew-self-adjoint (this is shown by other arguments in [R-S 1, p. 302]). Hence, the closure of $dU(A)$, which we also denote by $dU(A)$, is skew-self-adjoint and generates a strongly continuous one-parameter unitary group by (a transformation of) Stone's theorem. We denote this strongly continuous unitary one-parameter group by $U(tA)$. Notice that in the case of bounded A, the above M may be chosen equal to $\|A\|$ and $\mathcal{D}_A$ becomes all of $\mathcal{D}$ (all vectors in $\mathcal{H}$ is an analytic vector for A bounded).

The above mapping $A \to dU(A)$, which maps skew-self-adjoint operators on $\mathcal{H}$ into skew-self-adjoint operators on $\mathcal{F}_\vee(\mathcal{H})$ is called *the second quantization* mapping. Notice that we use the commutator $[B, C]$ to denote $BC - CB$, even for unbounded operators. Of course, one has to specify on which domain the commutator is defined. In our case the two operators have at least $\mathcal{D}$ as a common invariant domain, so the commutator is well-defined on $\mathcal{D}$. However, one has to be very careful, since a vanishing commutator on a dense set does not necessarily mean that the unbounded operators do commute (see for example [R-S 1, p. 272-273 and p. 306-307]).

Theorem 15 *The second quantization mapping, $A \to dU(A)$, on skew-self-adjoint operators A in $\mathcal{H}$ fulfils*

1) $U(tA)a(f)U(-tA) = a(e^{tA}f)$ *on $\mathcal{D}$, for analytic vectors f for A and $t \in \mathbf{R}$.*

2) $\overline{[dU(A), a^*(f)]} = a^*(Af)$ *for all $f \in \mathcal{D}(A)$.*

3) $dU([A, B]) = \overline{[dU(A), dU(B)]}$, *at least for A and B bounded and skew-self-adjoint on $\mathcal{H}$.*

Proof. 1)   The proof is similar to that of theorem 2 part 1), since for each analytic vector f_j for A, also $e^{-tA}f_j$ is an analytic vector for A. In addition, one have to control the domain, which is easily done.

2)   Follows by considerations completely analogical to those of theorem 2 part 2).

3)   In this case the proof of theorem 2 part 3) fails, however, the following proof could as well have been used in the earlier case,

theorem 2 part 3), with obvious modifications. For A and B bounded and skew-self-adjoint on $\mathcal{H}$

$$dU(A)dU(B)(f_1 \vee \ldots \vee f_n)$$

$$= \sum_{\substack{i,j=1 \\ j \neq i}}^{n} f_1 \vee \ldots \vee f_{i-1} \vee Bf_i \vee f_{i+1} \vee \ldots$$

$$\ldots \vee f_{j-1} \vee Af_j \vee f_{j+1} \vee \ldots \vee f_n$$

$$+ \sum_{i=1}^{n} f_1 \vee \ldots \vee ABf_i \vee \ldots f_n$$

and

$$[dU(A), dU(B)](f_1 \vee \ldots \vee f_n)$$

$$= \sum_{i=1}^{n} f_1 \vee \ldots \vee [A, B]f_i \vee \ldots \vee f_n$$

$$= dU([A, B])(f_1 \vee \ldots \vee f_n)$$

where $f_1 \vee \ldots \vee f_n \in V^n \mathcal{H}$ is chosen arbitrarily, hence

$$[dU(A), dU(B)] = dU([A, B])$$

on $\mathcal{D}$. □

Later, in section 3.3, we will use this result, but only for bounded operators A and B. Notice that part 3) of theorem 15 implies that the second quantization mapping $A \to dU(A)$ is a Lie algebra homomorphism.

A standard example arises for $A = i \cdot I$, then

$$dU(A)(f_1 \vee \ldots \vee f_n) = i \cdot n \cdot (f_1 \vee \ldots \vee f_n)$$

we call $N = -idU(A)$ for the number operator on $\mathcal{F}_V(\mathcal{H})$, with dense domain

$$\mathcal{D}(N) = \{F = \oplus_{n=0}^{\infty} F_n \in \mathcal{F}_V(\mathcal{H}) : \sum_{n=0}^{\infty} n^2 \cdot \|F_n\|^2 < \infty\}$$

So

$$N(\oplus_{n=0}^{\infty} F_n) = \oplus_{n=0}^{\infty}(n \cdot F_n)$$

for each $\oplus_{n=0}^{\infty} F_n \in \mathcal{D}(N)$. Since N is given in its spectral representation, it is evidently self-adjoint, by von Neumann's theorem (see
[R-S 1, p. 275]), as it should be, since $dU(A)$ is skew-self-adjoint.

In the next section we will use the method of the second quantization to construct a metapletic representation, i.e. a representation
of the restricted metaplectic group.

3.3 The Infinite-Dimensional Metaplectic Representation

In this section we introduce the symplectic group and construct
the so-called metaplectic representation of the restricted symplectic group on a Lie algebra level, by ideas similar to those of the
second quantization. We follow a strategy parallel to that outlined in section 2.4, in discussing the orthogonal group and the spin
representation, and use section 2.1 intensively.

Many authors have studied these subjects, however we will only
mention a few [Lu 2], [Ya], [P-S] and [Sh].

We use the notation hitherto used, exept $\mathcal{H}$ denotes both a complex Hilbert space, as usually, and the corresponding real Hilbert
space (if neccessary we write $\mathcal{H}_r$ in the latter case to avoid confusion). Moreover σ denotes the explicit non-degenerated symplectic
form, $\sigma(\cdot, \cdot) = Im\,\langle \cdot, \cdot \rangle$, on $\mathcal{H}_r$.

We define the symplectic group $Sp(\mathcal{H})$ as the group consisting of
real linear invertible operators $S : \mathcal{H} \to \mathcal{H}$ satisfying $\sigma(Sf, Sg) = \sigma(f, g)$, for all $f, g \in \mathcal{H}$. This definition is analogous to the definition
of the orthogonal group given in section 2.4. As in that case we may
introduce a complex structure in the real Hilbert space $\mathcal{H}_r$ (see also
the discussion in the end of section 3.1).

If $\{e_k\}_{k \in \mathbf{N}}$ is an orthonormal basis of $\mathcal{H}$ and we let u_k and v_k
denote e_k and $i \cdot e_k$, respectively, then $\{u_k, v_k\}_{k \in \mathbf{N}}$ form an orthonormal basis for $\mathcal{H}_r$ with respect to $\tau(\cdot, \cdot) = Re\,\langle \cdot, \cdot \rangle$ and the complex
structure of $\mathcal{H}$ is then reflected by a complex structure operator J
on $\mathcal{H}_r$, which is the real linear bounded operator, discussed in the
end of section 3.1, given by $Ju_k = v_k$ and $Jv_k = -u_k$, and fulfilling
$J^2 = -I$ and $\sigma(Jf, g) = -\sigma(f, Jg)$, for all $f, g \in \mathcal{H}_r$, i.e. $J^\sigma = -J$,
where J^σ denotes the transpose of J relative to $\sigma(\cdot, \cdot)$.

Because of the unique correspondence between $\mathcal{H}$ and $\mathcal{H}_r$, given
by $e_k \leftrightarrow u_k$ and $i \cdot e_k \leftrightarrow v_k = Ju_k$, we will not emphasize which space

we consider and therefore drop the r-index, unless confusion may arise. It now follows that any real linear mapping $S : \mathcal{H} \to \mathcal{H}$ can be split into a sum of a complex linear operator S_1 and a complex anti-linear operator S_2 as $S = S_1 + S_2$, by putting $S_1 = \frac{1}{2}(S - JSJ)$ and $S_2 = \frac{1}{2}(S + JSJ)$, as in the case of the orthogonal group. Notice that $[S_1, J] = 0$ and $[S_2, J]_+ = 0$. As in the earlier case, the subscripts 1 and 2 will refer to this splitting in the following. The adjoint S_1^* of S_1 is the usual adjoint of a complex linear operator, given by $\langle f, S_1 g \rangle = \langle S_1^* f, g \rangle$, for all $f, g \in \mathcal{H}$. In contrast, the adjoint S_2^* of the anti-linear operator S_2 is given by $\langle f, S_2 g \rangle = \langle g, S_2^* f \rangle$, for all $f, g \in \mathcal{H}$, as the usual adjoint of a complex linear operator S_2 from $\mathcal{H}$ to the conjugated Hilbert space $\mathcal{H}^*$. So the adjoint operation means different things, corresponding to the subscript of the operator it is applied to, even though we use an asterisk to denote it in both cases.

Since $S^\sigma S$ is the identity on $\mathcal{H}_r$, for any $S \in Sp(\mathcal{H})$, we get that $S^{-1} = S^\sigma$. Direct calculation gives

$$
\begin{aligned}
\sigma(S^\sigma f, g) &= \sigma(f, S_1 g) + \sigma(f, S_2 g) \\
&= \sigma(S_1^* f, g) + \sigma(g, S_2^* f) \\
&= \sigma((S_1^* - S_2^*) f, g)
\end{aligned}
$$

for all $f, g \in \mathcal{H}$, where we have used the skew-symmetry of the form $\sigma(\cdot, \cdot)$. Then it follows that

$$
S^{-1} = S^\sigma = S_1^* - S_2^*
$$

Notice that we then have $S_1^* = (S^\sigma)_1$, but $S_2^* = -(S^\sigma)_2$. Moreover

$$
\begin{aligned}
I &= S^{-1} S = S^\sigma S \\
&= (S_1^* - S_2^*)(S_1 + S_2) \\
&= (S_1^* S_1 - S_2^* S_2) + (S_1^* S_2 - S_2^* S_1)
\end{aligned}
$$

where each term is complex linear in the first bracket and complex anti-linear in the second bracket, so

$$
S_1^* S_1 - S_2^* S_2 = I \tag{3.17}
$$

and

$$
S_1^* S_2 - S_2^* S_1 = 0 \tag{3.18}
$$

Naturally $S_1 S_1^* - S_2 S_2^* = I$ and $S_2 S_1^* - S_1 S_2^* = 0$, in a similar manner. We will call S_1 and S_2 fulfilling these criteria, *Bogoliubov transformations*. From the last part of theorem 14 we know that such

Bogoliubov transformations determinate unique automorphisms on the CCR-algebra, by $W(f) \to W(S^{-1}f)$, for all $f \in \mathcal{H}$, where $W(f)$ denote the Weyl operators (we use the word automorphisms synonymously with *-automorphisms). Below we present an alternative proof, in a form to be used later.

Since $[\pi(S^{-1}f), \pi(S^{-1}g)] = \sigma(S^{-1}f, S^{-1}g) \cdot I = \sigma(f,g) \cdot I$ we observe that the commutation relations, on this form, are invariant under the action $\pi(f) \to \pi_S(f) = \pi(S^{-1}f)$, $S \in Sp(\mathcal{H})$. By the construction of the Weyl operators it follows that the Weyl forms of the canonical commutation relations are left invariant. As in the case of the orthogonal group, there is a natural question to ask; for which $S \in Sp(\mathcal{H})$ is the automorphism given by $\pi(f) \to \pi_S(f) = \pi(S^{-1}f)$, and then the automorphism given by $W(f) \to W_S(f) = W(S^{-1}f)$, unitary implementable (in the Fock representation), i.e. for which $S \in Sp(\mathcal{H})$ does there exist a unitary operator $U(S)$ on $\mathcal{F}_V(\mathcal{H})$ such that $\pi_S(f) = U(S)^*\pi(f)U(S)$, for all $f \in \mathcal{H}$, where $\pi_S(f) = \pi(S^{-1}f)$.

Before answering this question we have to define a subgroup $Sp_2(\mathcal{H})$, called *the restricted symplectic group*, of $Sp(\mathcal{H})$.

$$Sp_2(\mathcal{H}) = \{S \in Sp(\mathcal{H}) : S_2 \in \mathbf{L}_2(\mathcal{H})\}$$

where $\mathbf{L}_2(\mathcal{H})$, as before denotes the Hilbert-Schmidt operators on $\mathcal{H}$. It follows that $Sp_2(\mathcal{H})$ indeed is a group, since $(S^{-1})_2 = -S_2^*$ and $(ST)_2 = S_1T_2 + S_2T_1$ are Hilbert-Schmidt operators, for all $S, T \in Sp_2(\mathcal{H})$.

The restricted symplectic group can be given the structure of a topological group in several different ways, in complete analogy with the restricted orthogonal group. The verification in the case of the strongest topology, which is the topology given by the uniform topology on the linear part and the Hilbert-Schmidt topology on the anti-linear part, $\|S\|_{Sp_2} = \|S_1\| + \|S_2\|_{HS}$ is the same as in the case of the restricted orthogonal group, given in section 2.4 (except that we should write a Sp_2-index on the norm symbol instead of the $\mathcal{O}_2$-index). However, the restricted symplectic group is probably connected, but we are not aware of any proof of this.

The choice of topology on $Sp_2(\mathcal{H})$ determines the Lie algebra of $Sp_2(\mathcal{H})$. Our choice of "pre-Lie-algebra" is the simplest one, we choose

$$sp_2(\mathcal{H}) = \{A \in \mathbf{L}_r(\mathcal{H}) : A^\sigma = -A, A_2 \in \mathbf{L}_2(\mathcal{H})\}$$

where $L_r(\mathcal{H})$ denotes the bounded real linear operators on $\mathcal{H}$. The phrase "pre-Lie-algebra" means that we in some applications have to enlarge the "pre-Lie-algebra" to allow operators with unbounded linear parts. The demand $A^\sigma = -A$ means that $A_1^* = -A$, and $A_2^* = A_2$, where the adjoint is taken in their respective senses.

In the following we consider some particular S in $Sp_2(\mathcal{H})$ in a neighbourhood of the identity, generated from $sp_2(\mathcal{H})$ by the exponential mapping.

We return to the question: for which $S \in Sp(\mathcal{H})$ are the automorphisms, defined by $W(f) \to W_S(f) = W(S^{-1}f)$, unitarily implementable in the Fock representation, or equivalently, is there a unitary operator $U(S)$ on $\mathcal{F}_\vee(\mathcal{H})$ such that $\pi_S(f) = U(S)^{-1}\pi(f)U(S)$, for all $f \in \mathcal{H}$, where $\pi_S(f) = \pi(S^{-1}f)$. In fact, this question has already been answered by Shale in 1962, as stated in the following theorem.

Theorem 16 *A unitary operator $U(S)$, which implements the automorphism $\pi(f) \to \pi_S(f)$, exists if and only if $S \in Sp_2(\mathcal{H})$. Moreover, the operator $U(S)$ is unique up to a phase of modulus one.*

Proof. A proof can be found in [Sh], however, we make a construction of $U(S)$, for particular S in a neighbourhood of the identity in $Sp_2(\mathcal{H})$ below.   □

In the following we construct the unitary operator $U(S)$ for S in a neighbourhood of the identity in $Sp_2(\mathcal{H})$ in such a way that $U(\cdot)$ lifts one-parameter groups (near the identity) into one-parameter groups, whereby the phase is determined. We emphasize that this particular choice of fixing the phase is different from the choice made by Shale (see [Sh]). The following construction is largely analogous to the corresponding anti-symmetric case; however, there are some minor but essential differences. Below, we bring the details, pointing out the differences.

Since

$$U(TS)^{-1}\pi(f)U(TS) = (U(T)U(S))^{-1}\,\pi(f)\,(U(T)U(S))$$

for $S, T \in Sp_2(\mathcal{H})$, by direct calculation, and the *-algebra generated by $\{\pi(f) : f \in \mathcal{H}\}$ is irreducible, due to proposition 11, it follows that $U(T)U(S)U(TS)^{-1}$ is trivial, i.e. $U(T)U(S)U(TS)^{-1} = c(T, S)$.

I, where $c(T, S) \in$ C. Hence

$$U(T)U(S) = c(T, S) \cdot U(TS)$$

with $c(T, S) \in$ C. But the unitarity of $U(\cdot)$ forces $c(T, S)$ to be of modulus one, i.e. $|c(T, S)| = 1$. This means that the mapping $S \rightarrow U(S)$ defines a projective representation of the restricted symplectic group $Sp_2(\mathcal{H})$. The group cocycle $c(T, S)$ depends on the choice of the arbitrary phase in $U(S)$. It is possible to give an explicit formula for the cocycle $c(T, S)$, by choosing $U(S)$ such that $c(T, S)$ is smooth in such a way that $U(\cdot)$ lifts one-parameter groups into one-parameter groups, for S and T close to the identity, completely analogous to what we did in the orthogonal case (see section 2.4). We do this below, by first giving a constructive proof of the if-part of theorem 16, in the case of S in a neighbourhood of the identity in $Sp_2(\mathcal{H})$ of the form $S = e^A$, with $A \in sp_2(\mathcal{H})$. This is done by first constructing the metaplectic representation on a Lie algebra level, that is we construct $U(e^{sA})$, for $A \in sp_2(\mathcal{H})$ and $s \in$ R sufficiently small, by first constructing its skew-self-adjoint generator $dU(A)$, hence $U(e^{sA})$ is given by $e^{s \cdot dU(A)}$.

Consider first the complex linear part A_1 of $A \in sp_2(\mathcal{H})$, which is skew-self-adjoint. In this case $dU(A_1)$ and $U(e^{sA_1})$ are constructed by the method of Cook's second quantization given in section 3.2 (where we denoted $U(e^{sA_1})$ by $U(sA_1)$). Hence $dU(A_1)$ is skew-self-adjoint on $\mathcal{D}$ and $\mathcal{D}$ forms a dense set of analytic vectors for $dU(A_1)$, since A and then A_1 are bounded. Observe that $dU(A_1) : \vee^n \mathcal{H} \rightarrow \vee^n \mathcal{H}$.

We now turn to the anti-linear part A_2 of $A \in sp_2(\mathcal{H})$. A_2 is Hilbert-Schmidt and self-adjoint (in contrast to the orthogonal case). Due to the Hilbert-Schmidt condition of A_2, there exist two orthogonal sets $\{u_i\}_{i \in I}$ and $\{v_i\}_{i \in I}$ in $\mathcal{H}$, both spanning the range of A_2, such that A_2 has the representation

$$A_2 f = \sum_{i \in I} \langle f, v_i \rangle u_i$$

for any $f \in \mathcal{H}$. A direct calculation, using the definition of the adjoint of an anti-linear operator and the self-adjointness of A_2, gives

$$\sum_{i \in I} \langle f, v_i \rangle u_i = \sum_{i \in I} \langle f, u_i \rangle v_i$$

Notice that this equation differs from the analougous equation in the orthogonal case, by a change of sign. So we identify A_2 with a vector $A_2 \in \vee^2 \mathcal{H}$, where

$$A_2 = \sum_{i \in I} v_i \vee u_i = \sqrt{2} \cdot \sum_{i \in I} v_i \otimes u_i$$

Observe that $\left\| \frac{1}{\sqrt{2}} A_2 \right\|^2 = \left\| \sum_{i \in I} v_i \otimes u_i \right\| = \sum_{i \in I} \|v_i\|^2 \cdot \|u_i\|^2$.

The choice of A_2 ensures that the mapping $A_2 \mapsto \frac{1}{\sqrt{2}} A_2$ becomes an isometry with respect to the Hilbert-Schmidt norm on A_2. In fact, $\|A_2\|_{HS}^2 = \sum_{j \in \mathbf{N}} \|A_2 e_j\|^2$, where we may choose the orthonormal basis such that $e_i = \frac{v_i}{\|v_i\|}$ for $i \in I$, then

$$
\begin{aligned}
\|A_2\|_{HS}^2 &= \sum_{j \in \mathbf{N}} \left\| \sum_{i \in I} \langle e_j, v_i \rangle u_i \right\|^2 \\
&= \sum_{j \in \mathbf{N}} \sum_{i \in I} \delta_{i-j} \|v_i\|^2 \cdot \|u_i\|^2 = \sum_{i \in I} \|v_i\|^2 \cdot \|u_i\|^2
\end{aligned}
$$

Notice that the self-adjointness of A_2 mentioned above, is the reason why the above construction gives a vector $A_2 \in \vee^2 \mathcal{H}$. In the orthogonal case A_2 is skew-self-adjoint, reflecting the change of sign mentioned above, and the analogous construction gives a vector in $\wedge^2 \mathcal{H}$.

We are now able to map A_2 into an operator on $\mathcal{F}_\vee(\mathcal{H})$ by generalizing the idea of the creation operator, as we did in the orthogonal case. Define the operator $a^*(A_2)$ on product vectors by

$$a^*(A_2)\Omega = A_2 \tag{3.19}$$

and

$$a^*(A_2)(f_1 \vee \cdots \vee f_n) = A_2 \vee f_1 \vee \cdots \vee f_n \tag{3.20}$$

Notice that the vector $A_2 \vee f_1 \vee \cdots \vee f_n$ is well-defined in $\vee^{n+2}\mathcal{H}$ whenever $A_2 \in \vee^2 \mathcal{H}$ takes the form of a finite linear combination of the simple product vector, $A_2 = \sum_{i=1}^N g_{1,i} \vee g_{2,i} \in \vee^2 \mathcal{H}$, where $N \in \mathbf{N}$. Any vector $A_2 \in \vee^2 \mathcal{H}$ can be approximated by finite linear combinations of such product vectors. Let A_2 be an arbitrary vector in $\vee^2 \mathcal{H}$ and $A_2^{(N)} = \sum_{i=1}^N g_{1,i} \vee g_{2,i}$ an approximating sequence, i.e. $A_2 = \lim_{N \to \infty} A_2^{(N)}$, in $\vee^2 \mathcal{H}$. Then we may define $A_2 \vee f_1 \vee \cdots \vee f_n$ as the limit of $A_2^{(N)} \vee f_1 \vee \cdots \vee f_n$, in $\vee^{n+2}\mathcal{H}$, as N tends to infinity. The limit depends only on the vector $\mathcal{A}_2$ (and of course on $f_1 \vee \cdots \vee f_n$), but not on the choice of approximating sequence $\mathcal{A}_2^{(N)}$, since $\|a^*(\mathcal{A}_2)F_n\| \leq \sqrt{(n+2)(n+1)} \cdot \|\mathcal{A}_2\| \cdot \|F_n\|$, for $F_n \in \vee^n \mathcal{H}$. It follows that $a^*(\mathcal{A}_2)$ as an operator from $\vee^n \mathcal{H}$ into $\vee^{n+2}\mathcal{H}$ is bounded, for each $n \in \mathbb{N} \cup \{0\}$, by $\sqrt{(n+2)(n+1)} \cdot \|\mathcal{A}_2\|$. Even though the resulting bound is the same as in the anti-symmetric case, the calculations become a bit more technical in this case due to the actual symmetry. However, the proof is indeed similar to that in the anti-symmetric case (see section 2.4) and we omit the proof here. For a detailed proof see [Ya]. We notice that this upper bound is actually reached in certain cases, for example for $I = \{i_0\}$ and $F_n = (n!)^{-\frac{1}{2}} \cdot e_{i_0}^n$ is $\mathcal{A}_2 = \|\mathcal{A}_2\| \cdot e_{i_0}^2$ and

$$
\begin{aligned}
\|a^*(\mathcal{A}_2)F_n\|^2 &= \|\mathcal{A}_2\|^2 \cdot (n!)^{-1} \cdot \|e_{i_0}^{n+2}\|^2 = \|\mathcal{A}_2\|^2 \frac{(n+2)!}{n!} \\
&= (n+2)(n+1)\|\mathcal{A}_2\|^2 \cdot \|F_n\|^2
\end{aligned}
$$

Then we have the estimate

$$
\|a^*(\mathcal{A}_2)F_n\| \leq \sqrt{(n+2)(n+1)} \cdot \|\mathcal{A}_2\| \cdot \|F_n\|
$$

for $F_n \in \vee^n \mathcal{H}$. It is remarkable that we get the same bound as we did in the orthogonal case (of course the operators act in different spaces), considering that the creation and annihilation operators were bounded in contrast to the actual case. Of course $a^*(\mathcal{A}_2)$, as an operator on $\mathcal{F}_\vee(\mathcal{H})$, is unbounded (in the above example $\left\|a^*(\mathcal{A}_2)(n!)^{-\frac{1}{2}}e_{i_0}^n\right\| = \sqrt{(n+2)(n+1)} \cdot \|\mathcal{A}_2\|$, which tends to infinity as n).

It follows that $a^*(\mathcal{A}_2)$ defines a densely defined unbounded operator on $\mathcal{F}_\vee(\mathcal{H})$, with $\mathcal{D}$ as a dense set of analytic vectors, by completely similar reasons as those used in the anti-symmetric case (see section 2.4). Moreover, $a^*(\mathcal{A}_2) : \vee^n \mathcal{H} \to \vee^{n+2}\mathcal{H}$.

Consider $\langle a^*(f_1)F_n, \mathcal{A}_2 \rangle$, for arbitrary $f_1 \in \mathcal{H}$ and for any product vector $F_n \in \vee^n \mathcal{H}$, $n \in \mathbb{N} \cup \{0\}$. It is evidently zero for $n \neq 1$. For $n = 1$ we put $F = f_2$, whereby

$$
\begin{aligned}
\langle a^*(f_1)f_2, \mathcal{A}_2 \rangle &= \langle f_1 \vee f_2, \mathcal{A}_2 \rangle \\
&= \sum_{i \in I} (\langle f_1, v_i \rangle \langle f_2, u_i \rangle + \langle f_1, u_i \rangle \langle f_2, v_i \rangle) \\
&= \sum_{i \in I} \langle f_2, \langle f_1, v_i \rangle u_i + \langle f_1, u_i \rangle v_i \rangle \\
&= \langle f_2, 2\mathcal{A}_2 f_1 \rangle
\end{aligned}
$$

Since $\langle F, 2A_2 f_1 \rangle = 0$, for $F \notin \vee^1 \mathcal{H}$, we have

$$\langle a^*(f_1)F, A_2 \rangle = \langle F, 2A_2 f_1 \rangle$$

for all product vectors F, hence for all $F \in \mathcal{D}$. Then $A_2 \in \mathcal{D}(a(f))$ and $a(f)A_2 = 2A_2 f$. Thus

$$
\begin{aligned}
a(f)a^*(A_2)F_n &= a(f)(A_2 \vee F_n) \\
&= (a(f)A_2) \vee F_n + A_2 \vee (a(f)F_n) \\
&= (2a^*(A_2 f) + a^*(A_2)a(f))F_n
\end{aligned}
$$

for any $F_n = f_1 \vee \cdots \vee f_n \in \vee^n \mathcal{H}$, where we have used that $\mathcal{D}$ is invariant under both $a(f)$ and $a^*(A_2)$. Hence

$$\frac{1}{2} \cdot [a(f), a^*(A_2)] = a^*(A_2 f) \qquad (3.21)$$

on $\mathcal{D}$, for all $f \in \mathcal{H}$.

We denote the formal adjoint $a^*(A_2)^*$ of $a^*(A_2)$ by $a(A_2)$. In a similar manner as in the symmetric case it follows that $a(A_2)$ is well-defined on $\mathcal{D}$ and that it is inductively given on product vectors $f_1 \vee \cdots \vee f_n \in \mathcal{D}(a(A_2))$ by

$$a(A_2)(f_1 \vee \cdots \vee f_n) = (a^*(f_1)a(A_2) + 2a(A_2 f_1))(f_2 \vee \cdots \vee f_n)$$

and $a(A_2)\Omega = a(A_2)f = 0$, for $f \in \mathcal{D}(a(A_2))$. Observe that $a(A_2) : \vee^n \mathcal{H} \to \vee^{n-2} \mathcal{H}$, yields zero for $n = 0, 1$.

Moreover, it follows that $\mathcal{D}$ is a dense set of analytic vectors for $a(A_2)$, the proof of the anti-symmetric case carries over to the present case without problems.

Since $\mathcal{D}$ is invariant under the action of $a^*(f)$ the adjoint of formula (3.21) also holds on $\mathcal{D}$, i.e. $\frac{1}{2}[a(A_2), a^*(f)] = a(A_2 f)$, for all $f \in \mathcal{H}$. Thus $\mathcal{D}$ is invariant under the action of both $a^*(A_2)$ and $a(A_2)$.

Define $dU(A_2) = \frac{1}{2}(a(A_2) - a^*(A_2))$ on $\mathcal{D}$, for any self-adjoint anti-linear Hilbert-Schmidt operator A_2. Notice that $dU(A_2)$ is skew-symmetric by definition and that $dU(A_2) : \vee^n \mathcal{H} \to \vee^{n-2} \mathcal{H} \oplus \vee^{n+2} \mathcal{H}$. Define $dU(A) = dU(A_1) + dU(A_2)$ on $\mathcal{D}$, for any $A = A_1 + A_2 \in sp_2(\mathcal{H})$. Observe that $dU(A) : \vee^n \mathcal{H} \to \vee^{n-2} \mathcal{H} \oplus \vee^n \mathcal{H} \oplus \vee^{n+2} \mathcal{H}$

Theorem 17 *The operator $dU(A)$ is essentially skew-self-adjoint and fulfils*

$$[dU(A), \pi(f)] = \pi(Af) \qquad (3.22)$$

on $\mathcal{D}$ for $f \in \mathcal{H}$ and all $A \in sp_2(\mathcal{H})$,

$$\langle \Omega, dU(A)\Omega \rangle = 0 \tag{3.23}$$

for all $A \in sp_2(\mathcal{H})$, and

$$\langle \Omega, dU(A)dU(B)\Omega \rangle = -\frac{1}{4}\langle A_2, B_2 \rangle = -\frac{1}{2} \cdot \mathrm{Tr}(B_2 A_2) \tag{3.24}$$

Proof. Since $dU(A)$ is skew-symmetric and it follows that it has $\mathcal{D}$ as a dense set of analytic vectors, it follows that $dU(A)$ is essentially skew-self-adjoint, by a modification of Nelson's theorem ([R-S 2, p. 202], modified to skew-symmetric operators). From theorem 15 we have that $[dU(A_1), \pi(f)] = \pi(A_1 f)$ on $\mathcal{D}$, for all $f \in \mathcal{H}$ (since it holds for π replaced by a and a^* respectively).

For the anti-linear part of A we get, quite analogous to the proof of (2.9) in theorem 4, that

$$\begin{aligned}
[dU(A_2), \pi(f)] &= \frac{1}{2}[a(A_2), \pi(f)] - \frac{1}{2}[a^*(A_2), \pi(f)] \\
&= \pi(A_2 f)
\end{aligned}$$

and

$$\begin{aligned}
[dU(A), \pi(f)] &= [dU(A_1), \pi(f)] + [dU(A_2), \pi(f)] \\
&= \pi(A_1 f) + \pi(A_2 f) = \pi(Af)
\end{aligned}$$

on $\mathcal{D}$, for all $f \in \mathcal{H}$, proving (3.22). Moreover, since $dU(A_1)\Omega = 0$ and $a(A_2)\Omega = 0$, it follows that $\langle \Omega, dU(A)\Omega \rangle = -\frac{1}{2}\langle \Omega, a^*(A_2)\Omega \rangle = 0$, proving (3.23). Finally

$$\begin{aligned}
\langle \Omega, dU(A)dU(B)\Omega \rangle &= -\langle dU(A)\Omega, dU(B)\Omega \rangle \\
&= -\frac{1}{4}\langle a^*(A_2)\Omega, a^*(B_2)\Omega \rangle \\
&= -\frac{1}{4}\langle A_2, B_2 \rangle \\
&= -\frac{1}{2}\mathrm{Tr}(B_2, A_2)
\end{aligned}$$

on $\mathcal{D}$, where we have used the anti-symmetry of $dU(A)$ and (3.19). The spectral forms of A_2 and B_2, respectively, give the last equality as follows. Put

$$A_2 f = \sum_{i \in I} \langle f, v_i \rangle u_i = \sum_{i \in I} \langle f, u_i \rangle v_i$$

and

$$B_2 f = \sum_{j \in J} \langle f, x_j \rangle \, y_j = \sum_{j \in J} \langle f, y_j \rangle \, x_j$$

such that $A_2 = \sum_{i \in I} v_i \vee u_i$ and $B_2 = \sum_{j \in J} x_j \vee y_j$. Choose an orthonormal basis $\{e_k\}_{k \in \mathbb{N}}$ for $\mathcal{H}$ such that $e_i = \frac{v_i}{\|v_i\|}$, for $i \in I$. Then

$$
\begin{aligned}
\langle A_2, B_2 \rangle &= \sum_{i \in I} \sum_{j \in J} \langle v_i \vee u_i, x_j \vee y_j \rangle \\
&= \sum_{i \in I} \sum_{j \in J} \left(\langle v_i, x_j \rangle \langle u_i, y_j \rangle + \langle v_i, y_j \rangle \langle u_i, x_j \rangle \right) \\
&= \sum_{i \in I} \left(\left\langle u_i, \sum_{j \in J} \langle v_i, x_j \rangle \, y_j \right\rangle + \left\langle u_i, \sum_{j \in J} \langle v_i, y_j \rangle \, x_j \right\rangle \right) \\
&= 2 \cdot \sum_{i \in I} \langle u_i, B_2 v_i \rangle
\end{aligned}
$$

By $u_i = A_2 \frac{v_i}{\|v_i\|^2}$ and $A_2 e_k = 0$ for $k \notin I$, we get

$$
\begin{aligned}
\langle A_2, B_2 \rangle &= 2 \cdot \sum_{i \in I} \left\langle A_2 \frac{v_i}{\|v_i\|}, B_2 \frac{v_i}{\|v_i\|} \right\rangle \\
&= 2 \cdot \sum_{k \in \mathbb{N}} \langle A_2 e_k, B_2 e_k \rangle \\
&= 2 \cdot \sum_{k \in \mathbb{N}} \langle e_k, B_2 A_2 e_k \rangle \\
&= 2 \cdot \mathrm{Tr}(B_2 A_2)
\end{aligned}
$$

where we also have used the definition of the adjoint of an anti-linear operator B_2 together with the self-adjointness of B_2. Hereby equation (3.24) is proven. $\square$

We now define the unitary one-parameter group $U(e^{sA})$ by

$$U\left(e^{sA}\right) = e^{s \cdot dU(A)} \tag{3.25}$$

At this point we are rather close to having proved the if-part of theorem 16 for $S = e^{sA}$, $A \in sp_2(\mathcal{H})$, in a neighbourhood of the identity. However, the fact that $U(e^{sA})$ creates an infinite number of particles and that $\pi(f)$ is unbounded complicates the last part of the proof. Now, since

$$
\begin{aligned}
\frac{1}{t}\left(U(e^{tA}) - I\right)U\left(e^{tA}\right)F &= U\left(e^{sA}\right)\frac{1}{t}\left(U\left(e^{sA}\right) - I\right)F \\
&\rightarrow U\left(e^{sA}\right)dU(A)F
\end{aligned}
$$

for $F \in \mathcal{D}$, as $t \to 0$, it follows that the left-hand side also converges, so $U(e^{sA})F \in \mathcal{D}(dU(A))$ and $dU(A)U(e^{sA})F = U(e^{sA})dU(A)F$ (remembering that $dU(A)$ also denotes the skew-self-adjoint closure of the corresponding (preclosed) essentially skew-self-adjoint operator).

We will show that $U(e^{sA})\pi(f) = \pi(e^{sA}f)U(e^{-sA})$, however, the right-hand side is not immediately well-defined. But

$$
\pi(g)\sum_{n=0}^{m}\frac{s^n}{n!}dU(-A)^nF
$$

is well-defined for $F \in \mathcal{D}$ and $g \in \mathcal{H}$ arbitrary, since $dU(-A)\mathcal{D} \subset \mathcal{D}$. The series $\sum_{n=0}^{m}\frac{s^n}{n!}\pi(g)dU(-A)^nF$ converges in norm. In fact

$$
\|\pi(g)dU(-A)^nF\| \leq \sqrt{2 \cdot (N + 2n + 1)} \cdot \|g\| \cdot \|dU(-A)^nF\|
$$

by formula (3.10), where N has been chosen such that $F_k = 0$, for $k > N$ and we have used that $dU(-A)^nF \in \oplus_{l=0}^{N+2n}\vee^l\mathcal{H}$. Moreover, $\sum_{n=0}^{m}\frac{s^n}{n!}\|dU(-A)^nF\|$ converges for s sufficiently small, since $F \in \mathcal{D}$ is an analytic vector for $dU(-A)$, so

$$
\frac{s}{n+1} \cdot \frac{\|dU(-A)^{n+1}F\|}{\|dU(-A)^nF\|} \to 0 \quad \text{as } n \to \infty
$$

for $|s|$ less that some $s_0 \in \mathbf{R}_+$ (or $dU(-A)^nF = 0$ for n larger that some $N \in \mathbf{N}$). Then also

$$
\begin{aligned}
&\frac{s}{n+1} \cdot \sqrt{\frac{2(N + 2n + 3)}{2(N + 2n + 1)}} \cdot \frac{\|dU(-A)^{n+1}F\|}{\|dU(-A)^nF\|} \\
&= \frac{s}{n+1} \cdot \sqrt{1 + \frac{2}{N + 2n + 1}} \cdot \frac{\|dU(-A)^{n+1}F\|}{\|dU(-A)^nF\|}
\end{aligned}
$$

tends to zero as n tends to infinity for $|s|$ less that $s_0 \in \mathbf{R}_+$, giving the norm convergence of the series $\pi(g)\sum_{n=0}^{m}\frac{s^n}{n!}dU(-A)^nF$ for arbitrary

$g \in \mathcal{H}$, especially for $g = e^{-sA}f$. Hence by the closedness of $\pi(g)$

$$\lim_{m \to \infty} \sum_{n=0}^{m} \pi(g)\frac{s^n}{n!}dU(-A)^n F = \pi(g)\lim_{m \to \infty} \sum_{n=0}^{m} \frac{s^n}{n!}dU(-A)^n F$$

$$= \pi(g)U(e^{sA})F$$

proving that $\pi(g)U(e^{sA})$ is well-defined on $\mathcal{D}$ for all $g \in \mathcal{H}$, especially for g of the form $g = e^{sA}f$, where $f \in \mathcal{H}$. (Notice that $\pi(g)$ denotes the self-adjoint closure of the essentially self-adjoint operator $\frac{1}{\sqrt{2}}(a(g) + a^*(g))$).

To continue, we prove

$$dU(-A)^n\pi(f) = \sum_{k=0}^{n} \binom{n}{k}\pi((-A)^k f)dU(-A)^{n-k}$$

on $\mathcal{D}$, for all $n \in \mathbf{N}$ and $k = \{0, \ldots, n\}$, by induction. For $n = 0$ ($k = 0$), the formula is trivially true. For $n = 1$ ($k = 0, 1$) we obtain by (3.22), $dU(-A)\pi(f) = \pi(f)dU(-A) + \pi((-A)f)$, which clearly equals the right-hand side, as it should. Assume that the equation holds for a given $n \in \mathbf{N}$. Then

$$dU(-A)^{n+1}\pi(f) = dU(-A)\sum_{k=0}^{n}\binom{n}{k}\pi((-A)^k f)dU(-A)^{n-k}$$

$$= \sum_{k=0}^{n}\binom{n}{k}\pi((-A)^k f)dU(-A)^{n+1-k}$$

$$+ \sum_{k=0}^{n}\binom{n}{k}\pi((-A)^{k+1} f)dU(-A)^{n-k}$$

by the assumption and formula (3.22). By separating out the first term ($k = 0$) in the first sum, the last term ($k = n$) in the second sum and moreover making the transformation $k \to k - 1$ in the rest of the second sum, we get

$$dU(-A)^{n+1}\pi(f)$$

$$= \binom{n}{0}\pi((-A)^0 f)dU(-A)^{n+1} + \binom{n}{n}\pi((-A)^{n+1} f)dU(-A)^0$$

$$+ \sum_{k=1}^{n}\binom{n}{k}\pi((-A)^k f)dU(-A)^{n+1-k}$$

$$+ \sum_{k=1}^{n}\binom{n}{k-1}\pi((-A)^k f)dU(-A)^{n+1-k}$$

By the Pascal triangle (or direct calculation) $\binom{n}{k-1} + \binom{n}{k} = \binom{n+1}{k}$ and the fact that $\binom{n}{0} = \binom{n+1}{0} = \binom{n}{n} = \binom{n+1}{n+1} = 1$ we finally get the desired formula for $n+1$,

$$dU(-A)^{n+1}\pi(f) = \sum_{k=0}^{n+1} \binom{n+1}{k} \pi((-A)^k f) dU(-A)^{n+1-k}$$

and the induction is completed.

Using the derived formula we have

$$U\left(e^{sA}\right)\pi(f)F$$

$$= \lim_{m\to\infty} \sum_{n=0}^{m} \frac{s^n}{n!} dU(-A)^n \pi(f)F$$

$$= \lim_{m\to\infty} \sum_{n=0}^{m} \frac{s^n}{n!} \sum_{k=0}^{m} \binom{n}{k} \pi((-A)^k f) dU(-A)^{n-k} F$$

$$= \lim_{m\to\infty} \sum_{n=0}^{m} \frac{s^n}{n!} \frac{d^n}{ds^n}\bigg|_{s=0} \pi\left(e^{-sA}f\right) U\left(e^{-sA}\right) F$$

where we have used that $\pi(f)F \in \mathcal{D}$, for any $F \in \mathcal{D}$, and that $\pi(e^{-sA}f)U(e^{-sA})F$ is well-defined and C^∞ with respect to $s \in \mathbf{R}$ for any $F \in \mathcal{D}$. In fact

$$\frac{d^n}{ds^n}\pi\left(e^{-sA}f\right) U\left(e^{-sA}\right) F$$

$$= \sum_{k=0}^{n} \binom{n}{k} \pi\left((-A)^k e^{-sA} f\right) dU(-A)^{n-k} U\left(e^{-sA}\right) F$$

$$= \sum_{k=0}^{n} \binom{n}{k} \pi(g_k) U\left(e^{-sA}\right) \left(dU(-A)^{n-k}F\right)$$

is clearly well-defined inductively by the argument above and with $g_k = (-A)^k e^{-sA}f$, since $dU(-A)^{n-k}F \in \mathcal{D}$ for all $k \in \{0,\dots,n\}$. Reiterating the argument we obtain

$$U\left(e^{-sA}\right)\pi(f)F = \lim_{m\to\infty} \sum_{n=0}^{m} \frac{s^n}{n!} \frac{d^n}{ds^n}\bigg|_{s=0} \pi\left(e^{-sA}f\right) U\left(e^{-sA}\right) F$$

$$= \pi\left(e^{-sA}f\right) U\left(e^{-sA}\right) F$$

proving that $U(e^{-sA})\pi(f) = \pi(e^{-sA}f)U(e^{-sA})$ on $\mathcal{D}$, for all $f \in \mathcal{H}$, all $A \in sp_2(\mathcal{H})$ and s in a sufficiently small neighbourhood of zero.

We thereby get the formula

$$U(S)^{-1}\pi(f)U(S) = \pi(S^{-1}f) = \pi_S(f) \qquad (3.26)$$

on $\mathcal{D}$, for all $S = e^{sA}$, $A \in sp_2(\mathcal{H})$, $f \in \mathcal{H}$ and $s \in \mathbf{R}$ in a neighbourhood of zero, where $U(S)$ has been explicitely constructed, such that the arbitrary phase of $U(S)$ has been fixed on all one-parameter subgroups of $Sp_2(\mathcal{H})$ of the form $S = e^{sA}$, $A \in sp_2(\mathcal{H})$.

We call $U : S \rightarrow U(S)$ *the metaplectic representation* of the restricted symplectic group. It turns out that the elements $dU(A)$, $A \in sp_2(\mathcal{H})$, form a Lie algebra $mp_2(\mathcal{H})$, called *the metaplectic Lie algebra* corresponding to *the metaplectic group* $Mp_2(\mathcal{H})$ defined as the group of all unitary implementers $U(S)$, $S \in Sp_2(\mathcal{H})$ from Shale's theorem (theorem 16).

Theorem 18 *The bracket in the metaplectic Lie algebra is given by*

$$[dU(A), dU(B)] = dU([A, B]) + \omega(A, B) \cdot I \qquad (3.27)$$

on $\mathcal{D}$, for any $A, B \in sp_2(\mathcal{H})$, with the Lie algebra cocycle given by

$$\omega(A, B) = \frac{1}{2}\mathrm{Tr}([A_2, B_2])$$

$$= -\frac{i}{2}Im\langle A_2, B_2\rangle \qquad (3.28)$$

So $A \rightarrow dU(A)$ defines a projective representation of the Lie algebra $sp_2(\mathcal{H})$ of the restricted symplectic group onto the metaplectic Lie algebra $mp_2(\mathcal{H})$, in $\mathcal{F}_V(\mathcal{H})$.

Proof. From (3.26) and (3.11) it follows that

$$U(S^{-1})W(f)U(S) = W(S^{-1}f)$$

for all $f \in \mathcal{H}$ and $S \in Sp_2$. Then

$$U(e^{tA})U(e^{sB})U(e^{-tA})W(f)U(e^{tA})U(e^{sB})U(e^{-tA})$$
$$= W\left(e^{tA}e^{sB}e^{-tA}f\right)$$
$$= W\left(e^{s\cdot C(t)}\right)$$
$$= U\left(e^{s\cdot C(t)}\right)W(f)U\left(e^{-s\cdot C(t)}\right)$$

for all $f \in \mathcal{H}$, where $A, B, C(t) \in Sp_2(\mathcal{H})$ and $C(t)$ is given from A and B by the formula $C(t) = e^{tA}Be^{-tA}$, derived in the proof of theorem 2, part 3, of Chap. 2. Thus, the irreducability of the CCR-algebra, shown in corollary 12, gives

$$U(e^{tA})U(e^{sB})U(e^{-tA}) = b(tA, sB) \cdot U(e^{-s \cdot C(t)})$$

Of course this result also follows by consideration of the irreducability of *-algebra generated by $\{\pi(f) : f \in \mathcal{H}\}$ instead, using proposition 11. The unitarity of $U(\cdot)$ implies that $|b(tA, sB)| = 1$. Moreover, from the defining equation it follows directly that $b(0, sB) = b(tA, 0) = 1$. Now

$$b(tA, (s+r)B) \, U(e^{-(s+r)C(t)}) = b(tA, sB)b(tA, rB) \, U(e^{-(s+r)C(t)})$$

by the above and the one-parameter group properties. That is $s \rightarrow b(tA, sB)$ is a homomorphism. Furthermore $t \rightarrow b(tA, sB)$ is analytic. Then it follows that we may write

$$b(tA, sB) = e^{c(tA, sB)}$$

where $c(tA, 0) = c(0, sB) = 0$, $t \rightarrow c(tA, sB)$ is analytic and $c(tA, sB)$ is linear in s, so $c(tA, sB) = s \cdot c(tA, B)$, i.e.

$$b(tA, sB) = e^{s \cdot c(tA, B)}$$

Then

$$
\begin{aligned}
U(e^{tA})&dU(B)U(e^{tA}) \\
&= \left. \frac{d}{ds} \right|_{s=0} (U(e^{tA})U(e^{sB})U(e^{-tA})) \\
&= \left. \frac{d}{ds} \right|_{s=0} (e^{s \cdot c(tA,B)}U(e^{-s \cdot C(t)})) \\
&= c(tA, B) \cdot I + dU(C(t))
\end{aligned}
$$

on $\mathcal{D}$, and

$$
\begin{aligned}
U(e^{tA})&[dU(A), dU(B)]U(e^{-tA}) \\
&= \frac{d}{dt} (U(e^{tA})dU(B)U(e^{-tA})) \\
&= \frac{d}{dt} (c(tA, B) \cdot I + dU(C(t))) \\
&= c'(tA, B) \cdot I + dU(e^{tA}[A, B]e^{-tA})
\end{aligned}
$$

on $\mathcal{D}$, where we have used that $\frac{d}{dt}C(t) = e^{tA}[A, B]e^{-tA}$ and $c'(tA, B) = \frac{d}{dt}c(tA, B)$. Hence, for $t = 0$, $[d\tilde{U}(A), dU(B)] = dU([A, B]) + w(A, B) \cdot I$ on $\mathcal{D}$, where $w(A, B) = \frac{d}{dt}|_{t=0} c(tA, B)$.

Taking the vacuum expectation value of this equation we obtain

$$
\begin{aligned}
w(A, B) &= \langle \Omega, ([dU(A), dU(B)] - dU([A, B]))\Omega \rangle \\
&= \langle \Omega, dU(A)dU(B)\Omega \rangle - \langle \Omega, dU(B)dU(A)\Omega \rangle \\
&= -\frac{1}{2}\text{Tr}(B_2 A_2) + \frac{1}{2}\text{Tr}(A_2 B_2) \\
&= \frac{1}{2}\text{Tr}([A_2, B_2])
\end{aligned}
$$

by using equations (3.23) and (3.24). Since $\frac{1}{2}\text{Tr}(A_2 B_2) - \frac{1}{2}\text{Tr}(B_2 A_2) = \frac{1}{4}\langle B_2, A_2 \rangle - \frac{1}{4}\langle A_2, B_2 \rangle = -\frac{i}{2}Im\langle A_2, B_2 \rangle$, all claims of the theorem is proved. $\qquad\qquad\square$

We remark that the commutator of two anti-linear Hilbert-Scmidt operators does not, in general, vanish. Moreover, it follows directly from (3.27) that w is skew-symmetric and fulfils the Hochschild condition

$$
w(AB, C) + w(BC, A) + w(CA, B) = 0
$$

for all $A, B, C \in sp_2(\mathcal{H})$. Then the Jacobi identity is also fulfilled

$$
w([A, B], C) + w([B, C], A) = w([C, A], B) = 0
$$

for all $A, B, C \in sp_2(\mathcal{H})$, by which w is a closed two-form. In the special case, when the linear part of A and B are trace-class operators, we are able to transform the cocycle term away, by a change of phase, as follows. Put $dU_0(A) = dU(A) + \frac{1}{2}\text{Tr}(A_1) \cdot I$. Then a straight forward calculation gives

$$
\begin{aligned}
[dU_0(A), dU_0(B)] &= [dU(A), dU(B)] \\
&= dU([A, B]) + w(A, B) \cdot I \\
&= dU_0([A, B])
\end{aligned}
$$

since $w(A, B) = \frac{1}{2}\text{Tr}([A, B]_1)$ by

$$
\begin{aligned}
\text{Tr}([A, B]_1) &= \text{Tr}([A_1, B_1] + [A_2, B_2]) \\
&= 0 + \text{Tr}([A_2, B_2]) \\
&= 2 \cdot w(A, B)
\end{aligned}
$$

So we put $U_0(e^{sA}) = e^{s \cdot dU_0(A)} = e^{\frac{1}{2}s \cdot \text{Tr}(A_1)}U(e^{sA})$, for $s \in \mathbf{R}$, close to zero. Then

$$U(e^{sC}) = e^{\frac{1}{2}s(\text{Tr}(C_1)-\text{Tr}(A_1)-\text{Tr}(B_1))}U(e^{sA})U(e^{sB})$$

where C is given by the Campbell-Baker-Hausdorff formula, such that $e^{sC} = e^{sA}e^{sB}$ for s close to zero, i.e. the group cocycle $c(e^{sA}, e^{sB})$ is given by

$$c(e^{sA}, e^{sB}) = (\det(e^{sA_1}e^{sB_1}e^{-sC_1}))^{\frac{1}{2}}$$

for s close to zero, where $\det(e^D) = e^{\text{Tr}(D)}$ for any trace class operator D, by definition.

We conclude this section by calculating an explicit formula for *the vacuum functional*, given by $c(s) = \langle \Omega, U(e^{sA})\Omega \rangle$, for $A \in sp_2(\mathcal{H})$ and s in a neighbourhood of zero. Excactly as in the anti-symmetric case we arrive with

$$\Omega_s = c(s) \cdot e^{-\frac{1}{2}a^*(K)}\Omega$$

where $K = S_2 S_1^{-1}$ and $S = e^{sA}$ for s sufficiently small. This formula allows us to get a differential equation for $c(s)$, as follows

$$
\begin{aligned}
c'(s) &= \frac{d}{ds}\langle \Omega, \Omega_s \rangle = \langle \Omega, dU(A)\Omega_s \rangle \\
&= \langle -dU(A)\Omega, \Omega_s \rangle = \frac{1}{2}\langle a^*(A_2)\Omega, \Omega_s \rangle \\
&= \frac{1}{2}c(s)\left\langle A_2, e^{-\frac{1}{2}a^*(K)}\Omega \right\rangle = -\frac{1}{4}c(s)\langle A_2, K \rangle \\
&= -\frac{1}{2} \cdot \text{Tr}(KA_2) \cdot c(s)
\end{aligned}
$$

where we have used formula (3.25), the formula for Ω_s derived above, and the fact that $dU(A)\Omega = -\frac{1}{2}a^*(A_2)\Omega$. So $c(s)$ is determined by the differential equation above and the fact that $c(0) = \|\Omega\|^2 = 1$. Notice the change of sign relative to the corresponding anti-symmetric equation. Further, notice that $K = S_2 S_1^{-1} = (e^{sA})_2(e^{sA})_1^{-1}$ depends on $s \in \mathbf{R}$. Put $V_s = e^{-sA_1}(e^{sA})_1$ then

$$\frac{d}{ds}\text{Tr}(\log V_s) = \overline{\text{Tr}(KA_2)}$$

completely as in the anti-symmetric case. We may now rewrite the differential equation as

$$c'(s) = -\frac{1}{2}\frac{d}{ds}\overline{\text{Tr}(\log V_s)} \cdot c(s)$$

which has the solution

$$c(s) = \kappa \cdot e^{-\frac{1}{2}\overline{\mathrm{Tr}(\log V_s)}} = \kappa \cdot \left(\overline{\det(V_s)}\right)^{-\frac{1}{2}} = \kappa \cdot (\det(V_{-s}))^{-\frac{1}{2}}$$

since $c(-s) = \overline{c(s)}$.

In the previous paragraphs we have used that the determinant of V_s does exist, since $V_s - I = \int_0^s e^{-tA_1} A_2 (e^{tA})_2 \, dt$ is a trace-class operator, because both A_2 and $(e^{tA})_2$ are Hilbert-Schmidt operators (see for example [R-S 3, p. 322] or [Ar, p. 124]).

Finally, it follows that $\kappa = 1$, since $c(0) = 1$, hence

$$c(s) = (\det(V_{-s}))^{-\frac{1}{2}}$$

giving an explicit expression for the vacuum functional, as desired. Notice that this formula describes another power-law than in the corresponding expression for the spin representation. As a historical remark, we notice that $c(s) \neq 0$, so Shale's choice of fixing the phase by demanding $c(s) > 0$ was indeed possible (see [Sh, p. 157, theorem 4.1]). However, Shale's choice is different from ours which fixes the phase by demanding that $U(S)$ should be a one-parameter group if S is one. We summarize the above as follows.

Theorem 19 *The vacuum functional $c(s) = \langle \Omega, U(e^{sA})\Omega \rangle$ for $A \in sp_2(\mathcal{H})$ and s in a neighbourhood of zero, where $U(\cdot)$ denote the metaplectic representation, is simply*

$$c(s) = (\det(V_{-s}))^{-\frac{1}{2}}$$

where $V_{-s} = e^{sA_1}(e^{-sA})_1 = I - \int_0^s e^{tA_1} A_2 (e^{-tA})_2 \, dt$.

Proof. An immediately consequence of the above.   □

The metaplectic construction, derived in this chapter, will be used in Chap. 5 of applications, to construct some special representations of the diffeomorphism group $Diff^+(S^1)$ and of the Virasoro algebra in the symmetric Fock Hilbert space. But before doing so, we shall consider the loop algebras and the Virasoro algebra in general. Moreover, we will construct some special representations of the Virasoro in the anti-symmetric Fock Hilbert space.

Chapter 4

Loop Algebras and the Virasoro Algebra

4.1 Loop Algebras

In this section we introduce some basic theory of loop algebras and related subjects. The reason for doing so is partly to give a useful frame for the Virasoro algebra, which is considered in the next section, and partly to obtain knowledge on the loop group, which will be studied in a later chapter. We notice that the diffeomorphism group $Diff^+(S^1)$ acts as a group of automorphisms on loop groups and algebras and that the central extension of the Lie algebra $Vect(S^1)$ of $Diff^+(S^1)$ indeed is the Virasoro algebra. Moreover, the central extension of the loop algebra, together with the further extension discussed below, provide a nice example of Kac-Moody algebras, in fact, they are so-called affine Kac-Moody algebras (see [K-R,p.93-98], [P-S,p.76-78] and [Mi,p.21-23]). Futhermore, the simplest representation of the Kac-Moody algebras is given in terms of the spin representation, for details we refer to [Ar,p.124] (see also [Ve,p.1]). Hereby the connection between the spin representation and the loop algebra is clarified. There is an analogue connection between the metaplectic representation and the Virasoro algebra; however, the analogue construction is more cumbersome [Ve,p.1]. In Chap. 5 of applications we will show, in detail, how these subjects are related for some particular cases.

Since 1984 a huge number of papers have appeared on these subjects. The reason for this increasing interest is related to the important and succesful use of the theory in two-dimensional conformal theories and string theories in physics, see the examples in Chap. 1, introduction, or [F-H-J], [L-T], [Mi], [M-S] and [C-T].

This section is partly based on [P-S], [K-R], [Hu], [Mi], [G-W 1] and [G-W 2], as follows. The introduction of the loop algebras and the affine Kac-Moody algebras given below is a compound of the considerations given in [K-R, Chap. 9], [G-W 2, paragraph 1], [G-W 2, section 1] together with [Mi, Chap. 2]. The discussion concerning the Casimir operator and its generalization is a result of [P-S, section 9.4], [G-W 1, paragraph 2] and [K-R, Chap. 10]. Moreover, we will sometimes refer to results from [Hu].

Let $\mathfrak{g}$ be a finite dimensional simple Lie algebra over $\mathbf{C}$, i.e. it is a non-Abelian finite dimensional Lie algebra with no non-trivial ideals. We denote the Lie bracket on $\mathfrak{g}$ by $[\cdot, \cdot]_0$. Let $\mathbf{C}[t, t^{-1}]$ denote the ring of Laurent polynomials in $t \in \mathbf{C}$ and $t^{-1} \in \mathbf{C}$ and put

$$\tilde{\mathfrak{g}}_t = \mathfrak{g} \otimes \mathbf{C}[t, t^{-1}]$$

If $\mathbf{B} \subset \mathfrak{g}$ denotes a basis for $\mathfrak{g}$, then we denote the element $x \otimes t^n$ by $x(n)$, where $x \in \mathbf{B}$ and $t^n \in \mathbf{C}[t, t^{-1}]$, $n \in \mathbf{Z}$, is in the canonical basis for $\mathbf{C}[t, t^{-1}]$. Then

$$\{x(n) = x \otimes t^n : x \in \mathbf{B}, n \in \mathbf{Z}\}$$

form a basis for $\tilde{\mathfrak{g}}_t$. Notice the abuse of notation, since $x(n)$ is not the image of a point n under a mapping x. However, the used notation is quite common in this context.

We are mainly interested in the restriction from $t \in \mathbf{C} \backslash \{0\}$ to $t \in S^1 = \{z \in \mathbf{C} : |z| = 1\}$. In this case we sometimes write $t = e^{i\theta}$, where $\theta \in [0, 2\pi]$, and we write $\tilde{\mathfrak{g}}$ (without the index t) for the span of $\{x(n) = x \otimes e_n : x \in \mathbf{B}, n \in \mathbf{Z}\}$ where $e_n = e^{in\theta}$.

Sometimes one benefits from viewing $\tilde{\mathfrak{g}}$ as the space of smooth mappings $S^1\mathfrak{g}$, which is called *loops*, from S^1 to the Lie algebra $\mathfrak{g}$, i.e. we identify $\tilde{\mathfrak{g}}$ with $S^1\mathfrak{g}$. Notice that $S^1\mathfrak{g}$ is naturally an infinite-dimensional Lie algebra through the pointwise commutator

$$[x(m), y(n)]_\sim = [x, y]_0(m + n)$$

Here $S^1\mathfrak{g}$ is simply a vector space by pointwise addition of loops and natural multiplication of loops with scalars.

The above identification is, in fact, the reason for the name *loop algebra* for $\tilde{\mathfrak{g}} = S^1\mathfrak{g}$. Furthermore, we may identify $\tilde{f} \in \tilde{\mathfrak{g}}$ with $\sum_{n \in \mathbf{Z}} x_n e^{in\theta}$, where x_n in $\mathfrak{g}$ tends to zero as $|n|$ approach infinity, hence we may write $\tilde{f}$ in $\tilde{\mathfrak{g}}$ as a Fourier series with coefficients x_n in $\mathfrak{g}$. If $\mathfrak{g}$ is the Lie algebra of a group G, then $\tilde{\mathfrak{g}} = S^1\mathfrak{g}$ is the Lie algebra of the *loop group* LG, consisting of all smooth mappings from

S^1 into G (see for example [P-S, p. 27]). As above we may identify $\tilde{\mathfrak{g}}_t$ with the space of all formal Laurent mappings from $\mathbf{C} \setminus \{0\}$ to the Lie algebra $\mathfrak{g}$ and we may write an element $\tilde{f} \in \tilde{\mathfrak{g}}_t$ as

$$\tilde{f}(t) = \sum_{n \in \mathbf{Z}} x_n t^n$$

with $x_n \in \mathfrak{g}$.

Now let $(\cdot, \cdot)_0$ denote the Killing form on $\mathfrak{g}$, i.e. $(\cdot, \cdot)_0$ is the bilinear symmetric invariant form on $\mathfrak{g}$ given by

$$(x, y)_0 = \operatorname{Tr}(\operatorname{ad}(x) \circ \operatorname{ad}(y))$$

here $\operatorname{ad}(x)$ is given by $(\operatorname{ad}(x))(z) = [x, z]$, in the adjoint representation, for further details see for example [Mi, p. 4] or [Hu, p. 21].

Here invariance means

$$([x, y]_0, z)_0 = (x, [y, z]_0)_0$$

for all $x, y, z \in \mathfrak{g}$.

We may then define a skew-symmetric bilinear form ω on $\tilde{\mathfrak{g}}_t$ by

$$\omega(f, g) = \operatorname{res}_0 \left(\frac{df}{dt}, g \right)_0$$

for $f, g \in \tilde{\mathfrak{g}}_t$, where res_0 denotes the residual at zero. Then for arbitrary $x, y \in \mathfrak{g}$ we have

$$
\begin{aligned}
\omega(x(m), y(n)) &= m \cdot (x, y)_0 \cdot \operatorname{res}_0 t^{m+n-1} \\
&= m \cdot (x, y)_0 \cdot \delta_{m+n}
\end{aligned}
$$

Observe that ω is in fact skew symmetric and bilinear on $\tilde{\mathfrak{g}}_t$. Alternatively, for arbitrary $f, g \in \tilde{\mathfrak{g}}_t$, we may use the integral formula to calculate the residue as

$$\omega(f, g) = \frac{1}{2\pi i} \int_0^{2\pi} \left(\frac{df\left(e^{i\theta}\right)}{d\theta}, g\left(e^{i\theta}\right) \right)_0 d\theta$$

From the invariance and the symmetry of the Killing form $(\cdot, \cdot)_0$ on $\mathfrak{g}$, it follows that ω defines a two-cocycle, by use of the above formula

$$
\begin{aligned}
&\omega(x(m), [y(n), z(k)]_{\sim}) + \omega(y(n), [z(k), x(m)]_{\sim}) \\
&\quad + \omega(z(k), [x(m), y(n)]_{\sim}) \\
&= (m \cdot (x, [y, z]_0)_0 + n \cdot (y, [z, x]_0)_0 + k \cdot (z, [x, y]_0)_0) \cdot \delta_{m+n+k} \\
&= 0
\end{aligned}
$$

proving our claim. There is essentially no other cocycle on $\tilde{\mathfrak{g}}_t$ than ω
[P-S, p. 40]. The above construction also holds for the restriction of
t to S^1. Notice that the two-cocycle ω is invariant under the action
of the group $\text{Diff}^+(S^1)$ of orientation preserving diffeomorphisms of
the circle, as

$$\omega(f_\phi, g_\phi) = \omega(f, g)$$

where $f_\phi = f \circ \phi$ and $g_\phi = g \circ \phi$ for $\phi \in \text{Diff}^+(S^1)$. We return to a
discussion of $\text{Diff}^+(S^1)$ later on. However, this means that $\text{Diff}^+(S^1)$
acts as a group of automorphisms of the extended Lie algebra, which
we study below. Since

$$[\phi(x(m)), \phi(y(n))]_\sim = \phi([x(m), y(n)]_\sim)$$

by direct calculations, it follows that ϕ is a isomorphism and then
an automorphism of $S^1\mathfrak{g}$ because $\phi(x(m)) \in S^1\mathfrak{g}$.

Define $\hat{\mathfrak{g}}_t' = \tilde{\mathfrak{g}}_t \oplus \mathbf{C} \cdot c$ and $\hat{\mathfrak{g}}' = \tilde{\mathfrak{g}} \oplus \mathbf{C} \cdot c$ as the *central extensions*
of $\tilde{\mathfrak{g}}_t$ and $\tilde{\mathfrak{g}}$, respectively. Both are Lie algebras with respect to the
bracket or commutator given by

$$[f, g] = [f, g]_\sim + \omega(f, g) \cdot c$$
$$[f, c] = 0$$

for $f, g \in \tilde{\mathfrak{g}}_t$ and $f, g \in \tilde{\mathfrak{g}}$, respectively. The element c is, of course,
central. Expressed in terms of basis vectors this gives

$$[x(m), y(n)] = [x, y]_0(m + n) + m \cdot \delta_{m+n} \cdot (x, y)_0 \cdot c$$

for $x, y \in \mathfrak{g}$ and $m, n \in \mathbf{Z}$. Here $[x, y]_0(m+n)$ denotes $[x, y]_0 \otimes t^{m+n}$ by
our notational convention. The Lie algebras $\hat{\mathfrak{g}}_t'$ and $\hat{\mathfrak{g}}'$ are called *affine
Kac-Moody algebras* associated with $\mathfrak{g}$. Note that $\hat{\mathfrak{g}}_t'$ is generated by
$\{x(-1), x(0), x(1) : x \in \mathfrak{g}\}$ and that the mapping $x \mapsto x(0)$ is a
Lie algebra isomorphism from $\mathfrak{g}$ into $\hat{\mathfrak{g}}_t'$ i.e. it is invertible, linear
and conserves the commutator operation. Naturally the same holds
for $\hat{\mathfrak{g}}'$ instead of $\hat{\mathfrak{g}}_t'$, since the first is the restriction of the second to
$t \in S^1$.

We will identify $\mathfrak{g}$ with its image in $\hat{\mathfrak{g}}_t'$. We can now extend the
Killing form from $\mathfrak{g}$ to $\hat{\mathfrak{g}}_t'$ by defining

$$(f, g) = \text{res}_0 \left(t^{-1}(f(t), g(t))_0 \right)$$
$$= \frac{1}{2\pi} \int_0^{2\pi} \left(f\left(e^{i\theta}\right), g\left(e^{i\theta}\right) \right)_0 d\theta$$

and
$$(f, c) = (c, c) = 0$$

for $f, g \in \tilde{\mathfrak{g}}_t$. Notice that the form is indeed symmetric, invariant and bilinear, but degenerated due to $c \in \hat{\mathfrak{g}}'_t$. On basis vectors $x(m), y(n) \in \tilde{\mathfrak{g}}_t$ we obtain $(x(m), y(n)) = (x, y)_0 \cdot \delta_{m+n}$. Moreover, we may decompose $\hat{\mathfrak{g}}'_t$ as

$$\hat{\mathfrak{g}}'_t = \oplus_{n \in \mathbf{Z}} \, \mathfrak{g}(n) \oplus \mathbf{C} \cdot c$$

where $\mathfrak{g}(n)$ is spanned by $\{x(n) : x \in \mathfrak{g}\}$, identifying $\mathfrak{g}(0)$ with $\mathfrak{g}$, whereby $\hat{\mathfrak{g}}'_t$ is generated by $\mathfrak{g}(-1) \oplus \mathfrak{g}(0) \oplus \mathfrak{g}(1)$. Of course, we get the same structure if we use $\hat{\mathfrak{g}}'$ instead of $\hat{\mathfrak{g}}'_t$.

As mentioned above $\hat{\mathfrak{g}}'_t$ and $\hat{\mathfrak{g}}'$ are both degenerated with respect to the extended Killing form, due to the central element c. To avoid this inconvenience we will extend $\hat{\mathfrak{g}}'_t$ and $\hat{\mathfrak{g}}'$ once more. This is done in the following paragraphs.

Let $\mathfrak{h}$ be a Cartan subalgebra of $\mathfrak{g}$ (for further details see for example [Hu, p. 80] or [Mi, p. 5]), where Λ is the corresponding root system, Λ^+ is a system of positive roots, $\Lambda^- = -\Lambda^+$ implying $\Lambda = \Lambda^+ \cup \Lambda^-$ and $\Lambda_s \subset \Lambda^+$ is a systems of simple roots. The elements of Λ^+ can be written as a sum of elements from Λ_s. Now let $\mathfrak{g}_\alpha$ denote the (one-dimensional) root space corresponding to the root $\alpha \in \Lambda$. Choose $x_\alpha \in \mathfrak{g}_\alpha \setminus \{0\}$ and $y_\alpha \in \mathfrak{g}_{-\alpha} \setminus \{0\}$ for every $\alpha \in \Lambda^+$. The commutator on $\hat{\mathfrak{g}}'_t$ respectively $\hat{\mathfrak{g}}'$ then gives

$$\begin{aligned} [h, x_\alpha(n)] &= [h, x_\alpha]_0(0 + n) + n \cdot \delta_n \cdot (h, x_\alpha)_0 \cdot c \\ &= [h, x_\alpha]_0(n) \\ &= \alpha(h) x_\alpha(n) \end{aligned}$$

and similarly

$$\begin{aligned} [h, y_\alpha(n)] &= [h, y_\alpha]_0(n) \\ &= -\alpha(h) y_\alpha(n) \end{aligned}$$

We also have

$$[c, x_\alpha(n)] = [c, y_\alpha(n)] = [c, h] = 0$$

for $h \in \mathfrak{h}$. If we define a Cartan subalgebra of respectively $\hat{\mathfrak{g}}'_t$ and $\hat{\mathfrak{g}}'$ as $\mathfrak{h} \oplus \mathbf{C} \cdot c$, then each root α has infinite multiplicity. To avoid this, we extend $\hat{\mathfrak{g}}'_t$, respectively $\hat{\mathfrak{g}}'$, by joining an element d, to obtain

$$\hat{\mathfrak{g}}_t = \hat{\mathfrak{g}}'_t \oplus \mathbf{C} \cdot d = \tilde{\mathfrak{g}}_t \oplus \mathbf{C} \cdot c \oplus \mathbf{C} \cdot d$$

and the analogue without the t-index

$$\hat{\mathfrak{g}} = \hat{\mathfrak{g}}' \oplus \mathbf{C} \cdot d = \tilde{\mathfrak{g}} \oplus \mathbf{C} \cdot c \oplus \mathbf{C} \cdot d$$

Both extensions become Lie algebras with respect to the commutator given by the old one on $\hat{\mathfrak{g}}'_t$, respectively $\hat{\mathfrak{g}}'$, and in addition

$$[d, x(n)] = n \cdot x(n)$$

and

$$[d, c] = 0$$

It then follows that

$$[d, f(t)] = t\frac{df(t)}{dt}$$

for any $f \in \hat{\mathfrak{g}}'_t$ respectively $f \in \hat{\mathfrak{g}}'$, since this formula holds on basis vectors of the form $x(n)$. In the case of $\hat{\mathfrak{g}}$, the new element d has the following tangible realization, $d = -i\frac{d}{d\theta}$. This leads us to a definition of the Cartan algebra for $\hat{\mathfrak{g}}_t$, respectively $\hat{\mathfrak{g}}$, as

$$\hat{\mathfrak{h}} = \mathfrak{h} \oplus \mathbf{C} \cdot c \oplus \mathbf{C} \cdot d$$

We may write the roots in component form as $(\alpha, 0, n)$, correspon-ding to the above decomposition of $\hat{\mathfrak{h}}$. For example, $(\alpha, 0, n)$ is the root of the root vector $x_\alpha(n)$. So the above extension has reduced the root multiplicity to at most one, for $\alpha \in \Lambda \setminus \{0\}$.

For $\alpha = 0$, the roots $(0, 0, n)$ where $n \in \mathbf{Z} \setminus \{0\}$ have multiplicity $\dim(\mathfrak{h})$, since the vectors in $\mathfrak{h}(n)$ span the root subspace correspon-ding to the root $(0, 0, n)$. We choose to define the set of positive roots as

$$\hat{\Lambda}^+ = \{(\alpha, 0, n) : \alpha \in \Lambda, n \in \mathbf{N}\} \cup \{(\alpha, 0, 0) : \alpha \in \Lambda^+\}$$

and the set of negative roots as

$$\hat{\Lambda}^- = -\hat{\Lambda}^+$$

so that $\hat{\Lambda} = \hat{\Lambda}^+ \cup \hat{\Lambda}^-$ as in the case of finite dimensional (semi-) simple Lie algebras.

Proposition 20 *The affine Lie algebras $\hat{\mathfrak{g}}_t = \hat{\mathfrak{g}}'_t \oplus \mathbf{C} \cdot d$ and $\hat{\mathfrak{g}} = \hat{\mathfrak{g}}' \oplus \mathbf{C} \cdot d$ carries a non-degenerated, symmetric, invariant bilinear form $(\cdot, \cdot)$ given by*

$$
\begin{aligned}
(f(t), g(t)) &= \mathrm{res}_0 \left(t^{-1} \left(f(t), g(t) \right)_0 \right) \\
&= \frac{1}{2\pi} \int_0^{2\pi} \left(f(e^{i\theta}), g(e^{i\theta}) \right)_0 d\theta, \\
(f(t), c) &= (f(t), d) = (d, f(t)) \\
&= (c, c) = 0, \\
(c, d) &= 1
\end{aligned}
$$

and

$$
(d, d) = s
$$

for $f, g \in \tilde{\mathfrak{g}}_t$, respectively $f, g \in \tilde{\mathfrak{g}}$, and $s \in \mathbf{C}$.

Proof. This proposition is a generalization of [K-R, proposition 9.1], where $s = 0$. We extend their proof to the case of $s \neq 0$. We only prove the proposition for $\hat{\mathfrak{g}}_t$, since the proof for $\hat{\mathfrak{g}}$ is completely analogue. Notice that the form reduces to the old one when restricted to $\hat{\mathfrak{g}}'_t$. Since the old form was symmetric and bilinear the same properties evidently hold for the new form. However, the degeneracy of the old form is apparently eliminated by the demand $(c, d) = 1$. Consequently, the extended form is non-degenerate so the only thing left to prove is the invariance of the form. We know that the restriction to $\hat{\mathfrak{g}}'_t$ is invariant, and since $[c, d] = 0$ we just have to prove that

$$
([x(m), d], y(n)) = (x(m), [d, y(n)])
$$

This follows from a direct calculation:

$$
\begin{aligned}
([x(m), d], y(n)) &= (-m \cdot x(m), y(n)) \\
&= -m \cdot (x, y)_0 \cdot \delta_{m+n}
\end{aligned}
$$

and

$$
\begin{aligned}
(x(m), [d, y(n)]) &= (x(m), n \cdot y(n)) \\
&= n \cdot (x, y)_0 \cdot \delta_{m+n}
\end{aligned}
$$

hence, the propositon follows. $\square$

Corollary 21 *If $(\cdot, \cdot)$ is an invariant form on $\hat{\mathfrak{g}}_t$, then it is a positive multiple of the form defined in proposition 20.*

Proof. This corollary and a sketch of the proof can be found in [Mi, proposition 2.2.3]. We present the proof in detail. The invariance of the form yields

$$
\begin{aligned}
n \cdot (x_\alpha(n), x_\beta(m)) &= (([d, x_\alpha(n)], x_\beta(m)) \\
&= -((x_\alpha(n), [d, x_\beta(m)]) \\
&= -m \cdot (x_\alpha(n), x_\beta(m))
\end{aligned}
$$

so

$$
(n + m) \cdot (x_\alpha(n), x_\beta(m)) = 0
$$

i.e. $(x_\alpha(n), x_\beta(m)) = q_{\alpha\beta} \cdot \delta_{m+n}$, where $q_{\alpha\beta} = q_{\alpha\beta}(n)$ depends on n. If we choose an orthonormal basis with respect to $(\cdot, \cdot)_0$ instead, then the structure constants, given by $[x_\alpha, x_\beta]_0 = \sum_\mu a^\mu_{\alpha\beta} x_\mu$, obey

$$
\begin{aligned}
a^\gamma_{\alpha\beta} &= \sum_\mu a^\mu_{\alpha\beta}(x_\mu, x_\gamma)_0 \\
&= \sum_\mu a^\mu_{\alpha\gamma}(x_\alpha, x_\mu)_0 = a^\alpha_{\beta\gamma}
\end{aligned}
$$

Hence, it follows that

$$
a^\gamma_{\alpha\beta} = a^\alpha_{\beta\gamma} = a^\beta_{\gamma\alpha}
$$

Moreover, the anti-symmetry of the commutator gives $a^\gamma_{\alpha\beta} = -a^\gamma_{\beta\alpha}$, whereby we obtain the identities

$$
a^\gamma_{\alpha\beta} = a^\alpha_{\beta\gamma} = a^\beta_{\gamma\alpha} = -a^\gamma_{\beta\alpha} = -a^\beta_{\alpha\gamma} = -a^\alpha_{\gamma\beta}
$$

We shall especially use that $a^\gamma_{\alpha\beta} = -a^\gamma_{\beta\alpha}$ and $a^\gamma_{\alpha\beta} = -a^\gamma_{\beta\alpha}$. The invariance of the extended form yields

$$
\begin{aligned}
\sum_\mu q_{\alpha\mu} \cdot a^\mu_{\gamma\beta} &= \sum_\mu a^\mu_{\gamma\beta}(x_\alpha(n), x_\mu(-n)) \\
&= \sum_\mu a^\mu_{\alpha\gamma}(x_\mu(n), x_\beta(-n)) = \sum_\mu a^\mu_{\alpha\gamma} \cdot q_{\mu\beta}
\end{aligned}
$$

Then

$$
\sum_\mu q_{\alpha\mu} \cdot a^\beta_{\gamma\mu} = \sum_\mu a^\mu_{\gamma\alpha} \cdot q_{\mu\beta}
$$

since $a^\mu_{\gamma\beta} = -a^\beta_{\gamma\mu}$ and $a^\mu_{\alpha\gamma} = -a^\mu_{\gamma\alpha}$.

If we put $A_\gamma = \{a^\mu_{\gamma\alpha}\}_{\mu\alpha}$ and $Q = \{q_{\mu\beta}\}_{\mu\beta}$ then the above equation reads $QA_\gamma = A_\gamma Q$. Now, the adjoint representation with respect to the canonical basis $\{x_\alpha\}$ is represented by A_γ, since

$$(\mathrm{ad}\,(x_\gamma))\,(x_\beta) = [x_\gamma, x_\beta]_0 = \sum_\mu a^\mu_{\gamma\beta} x_\mu$$

Since the adjoint representation is irreducible for a simple Lie algebra and Q commutes with all A_γ it follows, by Schur's lemma (see for example [Hu, p. 26]) that Q is a multiple of the unit matrix I, i.e. $Q = \{q_{\alpha\beta}(n)\}_{\alpha\beta} = q(n) \cdot I$ giving

$$(x_\alpha(n), x_\beta(m)) = q(n) \cdot \delta_{n+m}\delta_{\alpha-\beta}$$

Further use of the invariance of the form gives

$$\begin{aligned}
a^\gamma_{\alpha\beta} \cdot q(n+1) &= \sum_\mu a^\mu_{\alpha\beta}\,(x_\mu(n+1), x_\gamma(-(n+1))) \\
&= a^\beta_{\gamma\alpha} \cdot q(n) = a^\gamma_{\alpha\beta} \cdot q(n)
\end{aligned}$$

where we repeatedly use

$$([x_\alpha(n), x_\beta(m)], x_\gamma(k)) = ([x_\alpha, x_\beta]_0\,(m+n), x_\gamma(k))$$

Consequently $q(n+1) = q(n) = q$. Hence, $(x_\alpha(n), x_\beta(m)) = q \cdot \delta_{n+m} \cdot \delta_{\alpha-\beta}$ and a renormalization of the form, by a constant factor, gives $(x_\alpha(n), x_\beta(m)) = \delta_{m+n} \cdot \delta_{\alpha-\beta}$, in accordance with the definition of the extended form in the proposition. Notice that

$$n \cdot (c, x_\alpha(n)) = (c, [d, x_\alpha(n)]) = ([c, d], x_\alpha(n)) = 0$$

and

$$n \cdot (d, x_\alpha(n)) = (d, [d, x_\alpha(n)]) = ([d, d], x_\alpha(n)) = 0$$

according to which $(c, f) = (d, f) = 0$ for any $f \in \tilde{\mathfrak{g}}_t$. $\qquad\square$

To complete this section on loop algebras we state some fundamental results, but first we define the Verma module for loop algebras. In fact what follows holds for arbitrary affine Lie algebras, denoted by $\hat{\mathfrak{g}}$.

We split $\hat{\mathfrak{g}}$ into subalgebras $\hat{\mathfrak{n}}_+, \hat{\mathfrak{n}}_-$ and $\hat{\mathfrak{h}}$ where $\hat{\mathfrak{n}}_+$ and $\hat{\mathfrak{n}}_-$ are spanned by the positive and negative roots, respectively, i.e.

$$\hat{\mathfrak{g}} = \hat{\mathfrak{n}}_- \oplus \hat{\mathfrak{h}} \oplus \hat{\mathfrak{n}}_+$$

where

$$\hat{\mathfrak{n}}_+ = \{x_\alpha(n) : \alpha \in \Lambda, n \in \mathbf{N}\} \cup \{x_\alpha \otimes 1 : \alpha \in \Lambda^+\}$$

and $\hat{\mathfrak{n}}_- = -\hat{\mathfrak{n}}_+$, which corresponds to a redefinition of the previous choice of negative and positive roots for the extended affine Kac-Moody algebra discussed earlier in connection with the loop algebras.

Let $\lambda \in \hat{\mathfrak{h}}^*$ be an arbitrary linear form on $\hat{\mathfrak{h}}$. We then define the Verma module as in the finite dimensional case, by

$$V_\lambda = U(\hat{\mathfrak{g}})/I_\lambda$$

where $U(\hat{\mathfrak{g}})$ denotes the universal enveloping algebra of $\hat{\mathfrak{g}}$ (see for example [Hu, p. 90]) and I_λ is the left ideal generated by $\hat{\mathfrak{n}}_+$ and the elements $h - \lambda(h)$, where $h \in \hat{\mathfrak{h}}$. Let $\mu \in \hat{\mathfrak{h}}^* \setminus \{0\}$, then it is a weight of V_λ if there exist a $v \in V_\lambda$ such that $h \cdot v = \mu(h) \cdot v$, for all $h \in \hat{\mathfrak{h}}$. If μ is a weight we define

$$V_\lambda(\mu) = \left\{ v \in V_\lambda : hv = \mu(h)v, \text{ for all } h \in \hat{\mathfrak{h}} \right\}$$

corresponding to the weight $\mu \in \hat{\mathfrak{h}}^* \setminus \{0\}$. We then have the following theorem.

Theorem 22 *The following holds for the Verma module V_λ of $\hat{\mathfrak{g}}$ (with the above notation):*

1) $\dim V_\lambda(\lambda) = 1$

2) *The set of weights μ of V_λ are of the form $\mu = \lambda - \sum_{i=1}^{l} r_i \alpha_i$ where $r_i \in \mathbf{N} \cup \{0\}$*

3) *V_λ is a direct sum of its weights subspaces $V_\lambda(\mu)$ and any invariant subspace respects this decomposition.*

4) $\dim V_\lambda(\mu) < \infty$

Proof. This theorem is a generalization of that in [Mi, theorem 1.5.4], where the Lie algebra considered is finite dimensional and semi-simple, and the proof given there is modified according to the present case. Let $v = 1 + I_\lambda \in V_\lambda$ and let ϕ denote the representation of $\hat{\mathfrak{g}}$ on V_λ given by the natural action, i.e. $\phi(x)(u + I_\lambda) = xu + I_\lambda$, for $x \in \hat{\mathfrak{g}}$ and $u \in U(\hat{\mathfrak{g}})$. Then $\phi(x_+)v = 0$ in V_λ, $\phi(h)v = \lambda(h) \cdot v$ in V_λ and $\phi(U(\hat{\mathfrak{g}}))v = V_\lambda$. So v is, by definition, a highest weight vector of weight λ and ϕ is the highest weight representation of $\hat{\mathfrak{g}}$ on V_λ, given by its natural action.

Now, let $\hat{\mathfrak{n}}_+$, $\hat{\mathfrak{n}}_-$ and $\hat{\mathfrak{h}}$ be spanned by x_{β_i}, y_{β_i}, where $i \in \mathbf{N}$, and h_i, $i = 1, \ldots, l$, respectively. Notice that the roots β_i are of the form $(\alpha, 0, n)$, $\alpha \in \Lambda^+$ and $n \in \mathbf{Z}$. From the Poincaré-Birkhoff-Witt theorem (see [Hu, p.92]) it follows that any ordered monomial in $U(\hat{\mathfrak{g}})$ can be written as a finite product of the form

$$u = \prod_i y_{\beta_i}^{r_i} \prod_j h_j^{s_j} \prod_k x_{\beta_k}^{p_k}$$

If any $p_k \neq 0$ then $uv = 0$, if any $s_j \neq 0$ then $h_j^{s_j} v = \lambda^{s_j} v$. So any element of V_λ can be written as a linear combination of vectors of the form

$$w = \prod_i y_{\beta_i}^{r_i} v$$

i.e. V_λ is generated by linear combinations of elements of the form $U(\hat{\mathfrak{n}}_-)v$. A direct calculation gives

$$h y_\beta^r = y_\beta^r (h - r \cdot \beta(h))$$

impying

$$
\begin{aligned}
hw &= h \prod_i y_{\beta_i}^{r_i} v \\
&= \left(\prod_i y_{\beta_i}^{r_i} \right) \left(h - \sum_i r_i \cdot \beta_i(h) \right) v \\
&= \left(\prod_i y_{\beta_i}^{r_i} \right) \left(\lambda(h) - \sum_i r_i \cdot \beta_i(h) \right) v \\
&= \left(\lambda(h) - \sum_i r_i \cdot \beta_i(h) \right) \left(\prod_i y_{\beta_i}^{r_i} v \right) \\
&= \mu(h) w
\end{aligned}
$$

Hence, each $w = \prod_i y_{\beta_i}^{r_i} v$ is an eigenvector of each $h \in \hat{\mathfrak{h}}$ with eigenvalue $\mu(h) = \lambda(h) - \sum_i r_i \cdot \beta_i(h)$, where $r_i \in \mathbf{N} \cup \{0\}$, proving part 2).

Notice that $\lambda = \mu$ if and only if all $r_i = 0$, whereby $V_\lambda(\lambda) = \mathrm{span}\{v\}$ and $\dim V_\lambda(\lambda) = 1$, giving the proof of part 1).

Moreover, each basis vector of the form $w = \prod_i y_\beta^{r_i} v$ in V_λ belongs to precisely one $V_\lambda(\mu)$, giving that $V_\lambda = \oplus_\mu V_\lambda(\mu)$, proving the first part of 3). Noticing that the number of vectors w, such that $\lambda - \sum_i r_i \cdot \beta_i$ equals a given μ, is finite and that these vectors span $V_\lambda(\mu)$, we observe that $\dim V_\lambda(\mu) < \infty$, proving part 4).

We now turn to the last part of 3). Let W be an invariant subspace of V_λ and choose an arbitrary $w \in W$. Then w can be written as a direct sum of finitely many elements $v_i \in V_\lambda(\mu_i) \setminus \{0\}$, with mutually different μ_i's, i.e.

$$w = \sum_{i=1}^m v_i \in \oplus_{i=1}^m V_\lambda(\mu_i)$$

Consequently $h^n w = \sum_{i=1}^m \mu_i^n v_i$ for each $n \in \mathbf{N} \cup \{0\}$ whence

$$\begin{pmatrix} w \\ hw \\ h^2 w \\ \vdots \\ h^{m-1} w \end{pmatrix} = \begin{pmatrix} 1 & 1 & \cdots & 1 \\ \mu_1 & \mu_2 & \cdots & \mu_m \\ \mu_1^2 & \mu_2^2 & \cdots & \mu_m^2 \\ \vdots & \vdots & \ddots & \vdots \\ \mu_1^{m-1} & \mu_2^{m-1} & \cdots & \mu_m^{m-1} \end{pmatrix} \begin{pmatrix} v_1 \\ v_2 \\ v_3 \\ \vdots \\ v_m \end{pmatrix}$$

Since the matrix is a Vandermonde matrix it is invertible, due to the fact that $\mu_i \neq \mu_j$, for $i \neq j$; in fact, its determinant is given by

$$\det\{\lambda_j^{i-1}\}_{i,j=1,\ldots,m} = \prod_{i>j}(\mu_i - \mu_j) \neq 0$$

The invariance of W under the action of $h \in \hat{\mathfrak{h}}$ and the invertibility of matrix $\{\lambda_j^{i-1}\}_{i,j=1,\ldots,m}$ then implies that each $v_i \in W$. Hence, $w \in \oplus_{i=1}^m (W \cap V_\lambda(\mu_i))$, i.e. $W = \oplus_\mu (W \cap V_\lambda(\mu))$, proving the remaining part of 3). $\qquad\square$

We now turn to the existence and uniqueness of irreducible highest weight representations of $\hat{\mathfrak{g}}$, in Verma modules.

Theorem 23 *The Verma module V_λ contains a unique maximal invariant proper submodule M_λ, moreover, $L_\lambda = V_\lambda/M_\lambda$ carries an irreducible highest weight representation of $\hat{\mathfrak{g}}$ with highest weight λ.*

Proof. This proof is identical to [Mi, theorem 2.4.1] and the proof is simply an expansion of that in [Mi]. Let M be any proper invariant subspace of V_λ. By the above theorem we may write $M = \oplus_\mu M(\mu)$, where $M(\mu) = M \cap V_\lambda(\mu)$. Then $M(\lambda) = \{0\}$, since otherwise the highest weight vector $v \in V_\lambda$ would belong to $M(\lambda)$ and $V_\lambda = U(\hat{\mathfrak{g}})v \subset M(\lambda) \subset M$, contradicting the property of M. Now, put M_λ equal to the union of all proper invariant subspaces of V_λ. Then M_λ becomes a maximal invariant subspace, by construction, and M_λ is proper, since it doesn't contain the highest weight vector $v \in V_\lambda$. Moreover M_λ is evidently unique. Therefore $L_\lambda = V_\lambda/M_\lambda$ contains no invariant proper subspaces and the representation of $\hat{\mathfrak{g}}$ on L_λ given by the natural action is obviously irreducible.

The highest weight vector in L_λ is given by $v_\lambda = v + M_\lambda$, since

$$h(v + M_\lambda) = \lambda(v + M_\lambda)$$

for any $h \in \hat{\mathfrak{h}}$,

$$\phi(x_+)(v + M_\lambda) = M_\lambda$$

for any $x_+ \in \hat{\mathfrak{n}}_+$, and

$$U(\hat{\mathfrak{g}})(v + M_\lambda) = V_\lambda + M_\lambda$$

where we have used that $v = 1 + I_\lambda$ is the highest weight vector for V_λ, with highest weight λ. Hence, λ is the highest weight and $v_\lambda = v + M_\lambda$ is the highest weight vector for $\hat{\mathfrak{g}}$ on L_λ. $\square$

Finally we will define the *Casimir operator* Ω for $\hat{\mathfrak{g}}$ as a generalization of, and by use of, the Casimir operator Ω_0 for the finite dimensional simple Lie algebras. Further generalizations of the Casimir operator Ω provide us with a highest weight representation of the Virasoro algebra through the Sugawara construction (see for example [G-W 1, p. 82]). The Virasoro algebra will be discussed in the next section.

Choose an orthonormal basis $\{x_i\}$ for the finite dimensional simple Lie algebra $\mathfrak{g}$ relative to $(\cdot, \cdot)_0$ and let λ_{ij} denote the structure coefficients of y and x_i i.e. $[y, x_i]_0 = \sum_j \lambda_{ij} x_j$ and $-\lambda_{ij} = \lambda_{ji}$ (summation over i's and j's without index sets here means summation over the finite basis index set). Let $y \in \mathfrak{g}$ and $m \in \mathbf{Z}$ be arbitrarily chosen and consider first $\sum_i x_i(-n)x_i(n)$, giving

$$[y(m), \sum_i x_i(-n)x_i(n)]$$

$$= \sum_i ([y, x_i]_0(m-n)x_i(n) + m \cdot \delta_{m-n}(y, x_i)_0 \cdot c \cdot x_i(n)$$

$$+ x_i(-n)[y, x_i]_0(m+n) + m \cdot \delta_{m+n}(y, x_i)_0 \cdot c \cdot x_i(-n))$$

$$= \sum_{i,j} \lambda_{ij} (x_j(m-n)x_i(+n) + x_i(-n)x_j(m+n))$$

$$+ \sum_i m \cdot \delta_{|m|-|n|} \cdot (y, x_i)_0 \cdot c \cdot x_i(m)$$

$$= \sum_{i,j} \lambda_{ij} (x_j(m-n)x_i(+n) - x_j(-n)x_i(m+n))$$

$$+ m \cdot y(m) \cdot c \cdot \delta_{|m|-|n|}$$

$$= X_{-n+m} - X_{-n} + m \cdot y(m) \cdot \delta_{|m|-|n|} \cdot c$$

where $X_n = \sum_{i,j} \lambda_{ij} x_j(n)x_i(m-n)$ and we have used

$$[y(m), x(n)] = [y, x]_0(m+n) + m \cdot \delta_{m+n} \cdot (y, x)_0 \cdot c$$

as given earlier, and that $y = \sum_i (y, x_i)x_i$. Define Ω_1 as the formal sum

$$\Omega_1 = -\sum_i \sum_{n \in N} x_i(-n)x_i(n) - \frac{1}{2}\sum_i (x_i(0))^2$$

Thus

$$[y(m), \Omega_1]$$

$$= -\sum_{n \in N} \left[y(m), \sum_i x_i(-n)x_i(n) \right] - \frac{1}{2} \left[y(m), \sum_i x_i(0)x_i(0) \right]$$

$$= -\sum_{n \in N} (X_{-n+m} - X_{-n}) - m \cdot y(m) \cdot c - \frac{1}{2}(X_m - X_0 + 0)$$

$$= -(\frac{1}{2}X_0 + X_1 + \cdots + X_{m-1} + \frac{1}{2}X_m) - m \cdot y(m) \cdot c$$

The skew-symmetry of λ_{ij} implies that

$$X_k + X_{m-k} = \sum_{i,j} \lambda_{ij} [x_j(k), x_i(m-k)]$$

$$= \sum_i [[y, x_i]_0, x_i]_0 \, (m)$$

$$= \left(\left(\sum_i (\mathrm{ad}(x_i))^2 \right) (y) \right) (m)$$

$$= (\Omega_0(y)) \, (m)$$

for any $k \in \{0, 1, \ldots, m\}$, where $\Omega_0 = \sum_i (\mathrm{ad}(x_i))^2$ denotes the Casimir operator in the adjoint representation of the finite dimensional simple Lie algebra $\mathfrak{g}$ (see for example [Hu, p. 27] or [G-W 1, p. 78]).

It is well known that Ω_0 commutes with every element of $\mathfrak{g}$, hence it belongs to the center of $\mathfrak{g}$ and therefore acts as a multiplication by a constant which we denote $2Q$ for later use. Additionally we note that

$$
\begin{aligned}
[y(m), \Omega_1] &= -\frac{1}{2} \sum_{k=0}^{m-1} (X_k + X_{m-k}) - m \cdot y(m) \cdot c \\
&= -m \left(\frac{1}{2} \Omega_0(y) \right) (m) - m \cdot y(m) \cdot c \\
&= -m(Q + c)y(m) \\
&= -(Q + c)[d, y(m)]
\end{aligned}
$$

since $[d, y(m)] = m \, y(m)$. Hence

$$[y(m), \Omega_1 - (Q + c)d] = 0$$

for all $y \in \mathfrak{g}$ and all $m \in \mathbf{Z}$, meaning that $\Omega = \Omega_1 - (Q + c)d$ commutes with all of $\hat{\mathfrak{g}}$. It is in this sense that the Casimir element Ω generalizes Ω_0. Observe that Ω is, in fact, independent of the choice of basis since Ω_0 and $\sum_i x_i(-n)x_i(n)$ are independent of this choice (see for example [P-S, p. 183]).

We will now generalize the Casimir operator defined above, but first we have to introduce what Kac (see [K-R, p. 107]) calls *an admissible representation*. We define a representation of $\hat{\mathfrak{g}}$ on a vector space V, or equivalently, a $\hat{\mathfrak{g}}$-module V, to be admissible if for all $x \in \mathfrak{g}$ and every vector $v \in V$ there exists a $N \in \mathbf{N}$, where $N = N(v)$, such that $(x(n))(v) = 0$, for $n > N$. For $x, y \in \mathfrak{g}$ and $k, n \in \mathbf{Z}$ we define the normal ordering $: x(k)y(n):$ of $x(k)$ and $y(n)$ (in agreement with, for example, [Mi, p. 174], [K-R, p. 108] and [G-W 1, p. 79]) by

$$
:x(k)y(n): = \begin{cases} x(k)y(n) & \text{for } k < n \\ \frac{1}{2}(x(k)y(n) + y(n)x(k)) & \text{for } k = n \\ y(n)x(k) & \text{for } k > n \end{cases}
$$

The following formal infinite sum

$$
T(n) = \frac{1}{2} \sum_{k \in \mathbf{Z}} \sum_i :x_i(k)x_i(n-k):
$$

for $n \in \mathbf{Z}$, reduces to a finite sum on any admissible space V, where $\{x_i\}$ denotes an orthonormal basis for $\mathfrak{g}$ with respect to $(\cdot, \cdot)_0$, $i = 1, \ldots, \dim \mathfrak{g}$. This is so since for any $m < 0$ there is only a finite number of terms $x_i(m)$ to the right due to the normal ordering, and given any $v \in V$ there is only a finite number of indices $m > 0$ such that $x_i(m)v \neq 0$, due to the admissibility condition. This also follows by the rewriting of $T(n)$, without normal ordering, as

$$
T(n) = \frac{1}{2} \sum_i \left(\sum_{k < \frac{1}{2}n} x_i(k)x_i(n-k) + \sum_{k \geq \frac{1}{2}n} x_i(n-k)x_i(k) \right)
$$

then for $k < n - N(v)$ the first term vanishes acting on $v \in V$ and for $k > N(v)$ the second term vanishes acting on $v \in V$, i.e.

$$
(T(n))(v)
$$

$$
= \frac{1}{2} \sum_i \left(\sum_{n-N \leq k < \frac{1}{2}n} x_i(k)x_i(n-k) + \sum_{\frac{1}{2}n \leq k \leq N} x_i(n-k)x_i(k) \right)(v)
$$

$$
= \left(\frac{1}{2} \sum_i \sum_{n-N \leq k \leq N} :x_i(k)x_i(n-k): \right)(v)
$$

Moreover, since $[x_i(k), x_i(n-k)] = [x_i, x_i]_0(n) = 0$ for $n \neq 0$ (because then the term $k \cdot \delta_{k+n-k} \cdot (x_i, x_i)_0 \cdot c$ vanishes), it follows that we may discard the normal ordering for $n \neq 0$. Moreover, since $k \in \mathbf{Z}$ we may also rewrite $T(n)$ as

$$
T(n) = \frac{1}{2} \sum_i \sum_{k \in \mathbf{Z}} :x_i(-k)x_i(n+k):
$$

Observe that $T(n)$ is homogeneous of degree n relative to d, i.e.

$$
[d, T(n)] = nT(n)
$$

when applied to a $v \in V$. This follows since $[d, x_i(m)] = m x_i(m)$ so

$$d x_i(k) x_i(m) = x_i(k)(d + k) x_i(m) = x_i(k) x_i(m)(d + m + k)$$

hence $[d, x_i(k) x_i(m)] = (k + m) x_i(k) x_i(m)$ and for $k + m = n$, the claim follows. Notice that $T(n)$ is independent of the choice of orthonormal basis, since so is $\sum_i x_i(k) x_i(n - k)$.

Since $T(0)$ is closely related to the Casimir operator, in fact, $-T(0)$ equals the previous defined operator Ω, we call the operators $T(n)$, $n \in \mathbf{Z}$, for the *shifted Casimir operators*, due to the following lemma:

Lemma 24 *For $y \in \mathfrak{g}$ and $m, n \in \mathbf{Z}$*

$$[y(m), T(n)] = m(Q + c) y(m + n)$$

as operators on V (admissible) where $2Q$ denotes the eigenvalue of the ordinary Casimir operator Ω_0 in the adjoint representation of $\mathfrak{g}$.

Proof. This lemma is almost identical to [G-W 1, lemma 2.1] and [K-R, proposition 10.1]. However, their proofs are diferent from ours. We use the same idea as we did when we considered the Casimir operator. But first, notice that the commutator of $: x_i(-k) x_i(n+k) :$ with any element $y(m)$ is independent of the normal ordering, since it only differs from $x_i(-k) x_i(n + k)$ by a constant multiplied by c, which commutes with $y(m)$. Furthermore, we observe that

$$
\begin{aligned}
\sum_i [[y, x_i]_0(p), x_i(q)] &= \sum_i [[y, x_i]_0, x_i]_0 (p + q) \\
&= (\Omega_0(y))(p + q) \\
&= 2 \cdot Q \cdot y(p + q)
\end{aligned}
$$

for any $p, q \in \mathbf{Z}$, since the term

$$p \cdot \delta_{p+q} ([y, x_i]_0, x_i)_0 \cdot c = p \cdot \delta_{p+q} (y, [x_i, x_i]_0)_0 \cdot c = 0$$

Due to the admissibility of the representation on V all infinite sums reduce to finite sums, when acting on V. Alternatively, one could use a cutoff procedure (see [K-R, p. 16]) and consider

$$T_\varepsilon(n) = \frac{1}{2} \sum_i \sum_{k \in \mathbf{Z}} : x_i(-k) x_i(n + k) : \chi(\varepsilon \cdot k)$$

which is a finite sum, since $\chi(x) = 1$ for $|x| \leq 1$ and $\chi(x) = 0$ for $|x| > 1$, and subsequently let ε approach zero. We use the first approach, though.

$$
\begin{aligned}
&[y(m), T(n)] \\
&= \left[y(m), \frac{1}{2}\sum_i \sum_{k \in \mathbf{Z}} :x_i(-k)x_i(n+k):\right] \\
&= \frac{1}{2}\sum_i \sum_{k \in \mathbf{Z}}\left([y, x_i]_0(m-k)x_i(n+k) + x_i(-k)[y, x_i]_0(m+n+k)\right) \\
&\quad + \frac{1}{2}\cdot m \cdot c \cdot \sum_{k \in \mathbf{Z}}\left(\delta_{m-k}\cdot\sum_i(y, x_i)_0 x_i(n+k) \right.\\
&\qquad\qquad\qquad\qquad \left. + \delta_{m+n+k}\cdot\sum_i(y, x_i)_0 x_i(-k)\right) \\
&= \frac{1}{2}\sum_i\left(\sum_{k\geq\frac{m-n}{2}} + \sum_{k<\frac{m-n}{2}}\right)[y, x_i]_0(m-k)x_i(n+k) \\
&\quad + \frac{1}{2}\sum_i\left(\sum_{k\geq-\frac{m+n}{2}} + \sum_{k<-\frac{m+n}{2}}\right)x_i(-k)[y, x_i]_0(m+n+k) \\
&\quad + \frac{1}{2}\cdot m \cdot c\,(y(n+m) + y(m+n))
\end{aligned}
$$

where the first sum term has been split into sums which are normally ordered (for $k \geq \frac{m-n}{2}$ respectively $k \geq -\frac{m+n}{2}$) and sums which are not normally ordered (for $k < \frac{m-n}{2}$ respectively $k < -\frac{m+n}{2}$). By use of the above commutator relations we turn the sums of the not normally ordered terms into sums of normally ordered terms. Then

$$
\begin{aligned}
[y(m), T(n)] &= \frac{1}{2}\sum_i \sum_{k\in\mathbf{Z}} :[y, x_i]_0(m-k)x_i(k+n): \\
&\quad + \frac{1}{2}\sum_i \sum_{k\in\mathbf{Z}} :x_i(-k)[y, x_i]_0(m+n+k): \\
&\quad + \sum_{k<\frac{m-n}{2}} Q\cdot y(m+n) - \sum_{k<-\frac{m+n}{2}} Q\cdot y(m+n) \\
&\quad + m\cdot y(n+m)\cdot c
\end{aligned}
$$

Observe that $Q\cdot y(m+n)$ is independent of k, so the additions with this term simply reduce to $m\cdot Q\cdot y(m+n)$. Hence, by the transformation $k \to k + m$, in the first addition we get

$$[y(m), T(n)] = \frac{1}{2} \sum_i \sum_{k \in \mathbf{Z}} :([y, x_i]_0(-k)x_i(k+m+n)$$
$$+ i(-k)[y, x_i]_0(k+m+n)):$$
$$+ m \cdot (Q + c) \cdot y(n+m)$$

Now, the terms in the addition cancel out when summing over i, since for any $y \in \mathfrak{g}$ and $p, q \in \mathbf{Z}$ we have

$$\sum_i x_i(p)[y, x_i]_0(q) = \sum_{i,j} x_i(p) \left([y, x_i]_0, x_j\right)_0 x_j(q)$$
$$= -\sum_{i,j} x_i(p) \left(x_i, [y, x_j]_0\right)_0 \cdot x_j(q)$$
$$= -\sum_j [y, x_j]_0(p)x_j(q)$$
$$= -\sum_i [y, x_i]_0(p)x_i(q)$$

where we have used the invariance of the Killing form. Finally we obtain the desired formula

$$[y(m), T(n)] = m \cdot (Q + c) \cdot y(n+m)$$

proving the lemma. $\square$

Theorem 25 *For $y \in \mathfrak{g}$ and $m, n \in \mathbf{Z}$*

$$[T(n), T(m)] = (Q+c) \cdot (n-m)T_{n+m} + \delta_{n+m} \cdot \frac{n^3 - n}{12} \cdot \dim \mathfrak{g} \cdot c \cdot (Q+c)$$

Proof. This theorem connects [K-R, theorem 10.1] and [G-W 1, lemma 2.2]. However, our proof utilizes the admissibility of the representation, which seems more appropriate in the present context. As in the above lemma, since we consider an admissible representation, the infinite sums appearing below become finite, when applied to any vector. It follows that

$$[T(n), T(m)] = \frac{1}{2} \sum_i \sum_{k \in \mathbf{Z}} [x_i(-k)x_i(k+n), T(m)]$$

$$= \frac{1}{2}\sum_i\sum_{k\in\mathbf{Z}} x_i(-k)(Q+c)(k+n)x_i(k+n+m)$$

$$+ \frac{1}{2}\sum_i\sum_{k\in\mathbf{Z}}(Q+c)(-k)x_i(-k+m)x_i(k+n)$$

Using the commutator relation to normally order the terms not arrangeed thus, i.e. for $k < -\frac{m+n}{2}$ in the first sum and for $k < \frac{m-n}{2}$ in the second sum, then we obtain

$$[T(n),T(m)]$$
$$= \frac{1}{2}(Q+c)\sum_i\sum_{k\in\mathbf{Z}}(k+n) : x_i(-k)x_i(k+n+m):$$

$$+ \frac{1}{2}(Q+c)\sum_{k<-\frac{m+n}{2}}(k+n)(-k)\cdot\delta_{-k+k+n+m}\cdot\dim\mathfrak{g}\cdot c$$

$$+ \frac{1}{2}(Q+c)\sum_i\sum_{k\in\mathbf{Z}}(-k) : x_i(-k+m)x_i(k+n):$$

$$+ \frac{1}{2}(Q+c)\sum_{k<\frac{m-n}{2}}(-k)(-k+m)\cdot\delta_{-k+m+k+n}\cdot\dim\mathfrak{g}\cdot c$$

Making the transformation $k \to k+m$ in the third sum gives

$$[T(n),T(m)] =$$
$$\frac{1}{2}(Q+c)\sum_i\sum_{k\in\mathbf{Z}}(n-m) : x_i(-k)x_i(k+n+m):$$

$$+ \frac{1}{2}(Q+c)\cdot c\cdot\dim\mathfrak{g}\cdot\delta_{n+m}\left(\sum_{k<0}-k(k+n)+\sum_{k<-n}-k(-k-n)\right)$$

Hence

$$[T(n),T(m)] = (Q+c)(n-m)T_{n+m}+\delta_{n+m}\frac{n^3-n}{12}\cdot\dim\mathfrak{g}\cdot c\cdot(Q+c)$$

where we have used that the last sums reduce to $\sum_{k=1}^{n-1}k(n-k) = \frac{n^3-n}{6}$ proving the theorem. $\square$

In any representation the operator c acts as multiplication by a scalar which we also denote c. Let us consider any admissible representation. If $c \neq -Q$ it is customary to consider

$$L_n = \frac{1}{(Q+c)}\cdot T(n)$$

instead of $T(n)$. The L_n, $n \in \mathbf{Z}$, then fulfils, the commutation relations

$$[L_n, L_m] = (n - m) \cdot L_{n+m} + \delta_{n+m} \cdot \frac{n^3 - n}{12} \cdot \frac{c \cdot \dim \mathfrak{g}}{Q + c}$$

and

$$[L_n, x(m)] = -m \cdot x(m + n)$$

for all $m \in \mathbf{Z}$ and $x \in \mathfrak{g}$. Beware that Q depends on $\mathfrak{g}$ (see [K-R, p. 111]). The construction above is known as *the Sugawara construction* and, as we will see in the next section, it provides a representation of the so-called Virasoro algebra (since it fulfils the commutation relations of the Virasoro algebra).

The loop algebra and the loop group have been studied intensively in the last few years, and there is a lot more to be said about these subjects, than we will do here. For further reading we refer to [Jø], [Mi], [P-S], [G-W 1], [G-W 2] and [K-R]. However, we notice that the algebra of smooth vector fields on the circle S^1 has a natural action as derivatives on the loop algebra, and the central extension of the algebra of smooth vector fields on the circle S^1 is, indeed, the above mentioned Virasoro algebra, which we will consider in the following section.

4.2 The Virasoro Algebra

The Virasoro algebra was introduced by the physicist M.A. Virasoro in 1970 ([Vi]). It is a relatively well-behaved infinite dimensional Lie algebra, which can be obtained as the central extension of the complexification of the smooth real vector fields on the unit circle S^1. Hence, the Virasoro algebra can be viewed as the central extension of the complexification of the Lie algebra of the diffeomorphism group. However, the exponential mapping from the Lie algebra of real smooth vector fields on the unit circle to the diffeomorphism group is neither locally one-to-one nor onto.

It turns out that the orientation preserving diffeomorphism group on the unit circle $Diff^+(S^1)$ acts as a group of automorphisms on any loop group and that the orientation preserving subgroup $Diff^+(S^1)$ acts projectively on all the known representations of the loop groups (see [P-S, p. 5]). Furthermore, we mention that the Virasoro algebra plays an important role in theoretical physics, e.g. conformal field theory (see [B-P-Z 1]), string theory (see [C-T] and [L-T]) and statistical physics (see [B-P-Z 2]). We refer to the physical motivation given in the introduction for further details and return to some of the above questions in the end of this section.

The relation of the Virasoro algebra to the spin representation and the metaplectic representation will be shown for some particular cases in Chap. 5 of applications.

In this section we introduce and discuss the Virasoro algebra, with its applications in mind. In addition, we will, construct some representations of the Virasoro algebra, in the anti-symmetric Fock Hilbert space.

Let $Vect(S^1)$ be the set of all smooth real vector fields on the circle S^1, i.e. $X \in Vect(S^1)$ has, in local coordinates, the form $X(\theta) = f(\theta)\frac{d}{d\theta}$ where f is a smooth real valued function on S^1 with period 2π. $Vect(S^1)$ is organized as a vector space by pointwise addition and natural multiplication by scalars (from $\mathbf{R}$). A direct calculation shows that the Lie bracket, also called the commutator, given by

$$[f(\theta)\frac{d}{d\theta}, h(\theta)\frac{d}{d\theta}] = (f(\theta)h'(\theta) - f'(\theta)h(\theta))\frac{d}{d\theta}$$

where $f'(\theta) = \frac{d}{d\theta}f(\theta)$ and $h'(\theta) = \frac{d}{d\theta}h(\theta)$, turns $Vect(S^1)$ into a Lie algebra (it is only the Jacobi identity which is not completely trivial to verify).

A smooth function f on S^1 is always square integrable and a basis can be taken as $\{1, \cos n\theta, \sin n\theta\}_{n\in\mathbf{N}}$. Thus, a basis for $Vect(S^1)$ over $\mathbf{R}$ is given by the vector fields $\frac{d}{d\theta}, \cos(n\theta)\frac{d}{d\theta}$ and $\sin(n\theta)\frac{d}{d\theta}$, $n \in \mathbf{N}$. Instead of considering the Lie algebra of smooth real vector fields on the circle S^1, we could consider the Lie algebra of smooth complex vector fields on the circle S^1, with basis given by the vector fields $e^{in\theta}\frac{d}{d\theta}$, $n \in \mathbf{Z}$.

This complex Lie algebra is in accordance with the complexification of $Vect(S^1)$, the complex linear span of (basis) vectors from $Vect(S^1)$. We denote this complex Lie algebra by $\mathfrak{d}$. It follows that we may consider $d_n = i \cdot e^{in\theta}\frac{d}{d\theta} = e^{in\theta}d_0$ as a basis for $\mathfrak{d}$. These basis elements satisfy the following commutation relations

$$[d_m, d_n] = (m - n) \cdot d_{m+n}, \quad m, n \in \mathbf{Z}$$

The Lie algebra $Vect(S^1)$ can be considered as the Lie algebra of the group G of orientation preserving (real) diffeomorphisms of S^1, i.e.

$$G = Diff^+(S^1) = \left\{ \gamma : \gamma(e^{i\theta}) = e^{i\phi(\theta)}, \quad \text{for some } \phi \in \mathcal{M} \right\}$$

where $\mathcal{M}$ is the set of smooth real 2π-periodic functions ϕ, for which $\phi'(\theta) > 0$.

The product of two elements $\xi, \eta \in G$ is given by the composition of mappings $(\xi, \eta)(z) = \xi(\eta(z))$ for each $z = e^{i\theta} \in S^1$. We can define a representation π of G on the vector space of smooth complex valued functions f on S^1 by

$$\pi(\gamma)f(z) = f(\gamma^{-1}(z))$$

It is clearly a representation, since

$$
\begin{aligned}
\pi(\gamma_1 \circ \gamma_2)f(z) &= f\left(\gamma_2^{-1}\left(\gamma_1^{-1}(z)\right)\right) \\
&= \left(f \circ \gamma_2^{-1}\right)\left(\gamma_1^{-1}(z)\right) \\
&= \left(\pi(\gamma_1)\left(f \circ \gamma_2^{-1}\right)\right)(z) \\
&= \pi(\gamma_1)\left(\pi(\gamma_2)f\right)(z) \\
&= \left(\pi(\gamma_1)\pi(\gamma_2)\right)f(z)
\end{aligned}
$$

notice that $\pi(\gamma)$ is invertible for every $\gamma \in G$, since $\gamma^{-1} \in G$ and $\pi(\gamma)^{-1} = \pi(\gamma^{-1})$. We may write the d_n's in the z-coordinate, $z = e^{i\theta}$, as $d_n = -z^{n+1}\frac{d}{dz}$, $n \in \mathbf{Z}$. The Fourier expansion then gives

$$\epsilon(z) = \gamma(z) - z = \sum_{n \in \mathbf{Z}} \epsilon_n z^{n+1}$$

where ϵ_n denotes the $(n+1)$'th Fourier coefficient of $\epsilon(z)$. Normally one take γ close to the identity, whence

$$\gamma^{-1}(z) \simeq z - \sum_{n \in \mathbf{Z}} \epsilon_n z^{n+1}$$

where $\simeq$ means equal up to first order in the ϵ_n's. Then

$$\pi(\gamma)f(z) \simeq f\left(z - \sum_{n \in \mathbf{Z}} \epsilon_n z^{n+1}\right)$$

$$\simeq\ f(z) - \sum_{n\in\mathbf{Z}}\epsilon_n z^{n+1}\frac{d}{dz}f(z)$$

$$= \left(1 + \sum_{n\in\mathbf{Z}}\epsilon_n d_n\right)f(z)$$

This shows that the d_n's form a basis of the complexification of the Lie algebra of the group G.

In the following we will consider $Vect(S^1)$ as a subalgebra over $\mathbf{R}$, consisting of real elements of the complex Lie algebra $\mathfrak{d}$. The real elements of $\mathfrak{d}$ are those which are skew-invariant under the complex conjugation, which maps d_n into d_{-n}. That is, defining an anti-linear anti-involution α on $\mathfrak{d}$ by

$$\alpha(d_n) = d_{-n}$$

and

$$\alpha\left(\sum\lambda_n d_n\right) = \sum\bar{\lambda}\alpha(d_n)$$

where $\lambda_n\in\mathbf{C}$, then

$$\begin{aligned}
\alpha\left([d_m, d_n]\right) &= (m-n)d_{-m-n}\\
&= -[d_{-m}, d_{-n}]\\
&= [\alpha(d_n), \alpha(d_m)]
\end{aligned}$$

for all $x, y\in\mathfrak{d}$. Hence, $Vect(S^1)\subset\mathfrak{d}$ consists of elements invariant under the action of $-\alpha$.

Let V be an arbitrary vector space with an Hermitian form $\langle\cdot,\cdot\rangle$, which, in our case, almost always is an inner product. Consider a unitary representation of $G = Diff^+(S^1)$ on V. If we suppress the representation symbol and identify the elements in G with the corresponding operator on V, that is, we organize V as a G-module, then

$$\langle\gamma(u), \gamma(v)\rangle = \langle u, v\rangle$$

for any $u, v\in V$ and $\gamma\in G$. On the Lie algebra level, this implies

$$\langle x(u), v\rangle = -\langle u, x(v)\rangle$$

for $x\in Vect(S^1)$. Since $x = -\alpha(x)$ for any $x\in Vect(S^1)$, by construction of α, we see that

$$\langle x(u), v\rangle = \langle u, \alpha(x)(v)\rangle$$

for any $x \in Vect(S^1)$. In general we say that a Hermitian form $\langle \cdot, \cdot \rangle$ is contravariant if $\langle x(u), v \rangle = \langle u, \alpha(x)(v) \rangle$ for all x in a Lie algebra $\mathfrak{g}$ and all u and v in a representation space of $\mathfrak{g}$, where α is an anti-linear anti-involution on $\mathfrak{g}$.

In the case where the form is non-degenerate, this means that $x^* = \alpha(x)$ for all $x \in \mathfrak{g}$, where x^* denotes the Hermitian adjoint with respect to the non-degenerate Hermitian form. We call the representation unitary if the contravariant Hermitian form, in addition, is positive definite.

In [K-R, p. 7] concrete irreducible representations of G on the density spaces are constructed.

It appears that some of these density spaces carry non-degenerate Hermitian contravariant forms and that all these representations are unitary. We will not make this construction here (a detailed treatment of this is given in [K-R, p. 7]), but rather consider the question of a central extension of $\mathfrak{d}$, yielding the Virasoro algebra.

The Lie algebra $\mathfrak{d}$ admits a non-trivial two-cocycle ω, namely

$$w(d_m, d_n) = \delta_{m+n} \cdot \frac{m(m^2 - 1)}{12}$$

where ω has been normalized to vanish on the $\mathfrak{sl}_2$-subalgebra spanned by $\{d_{-1}, d_0, d_1\}$. (The factor $\frac{1}{12}$ is a convention, see below).

By a direct calculation, the two-cocycle property follows:

$$w(d_k, [d_m, d_n]) + w(d_m, [d_n, d_k]) + w(d_n, [d_k, d_m]) = 0$$

In general for $X = f(e^{i\theta})\frac{d}{d\theta}$ and $Y = h(e^{i\theta})\frac{d}{d\theta}$ it reads

$$w(X, Y) = \frac{1}{24\pi i} \int_0^{2\pi} \left(f''(e^{i\theta}) + f(e^{i\theta}) \right) h'(e^{i\theta}) d\theta$$

as a direct calculation shows.

Given any two-cocycle ω, we may define a central extension of the Lie algebra $\mathfrak{d}$ as follows. The central extension $\mathfrak{d}_c = \mathfrak{d} \oplus \mathbf{C} \cdot c$ of $\mathfrak{g}$ is spanned by $\mathfrak{d}$ on some additional element, usually denoted by c and called the central charge, satisfying

$$[d_n, c]_c = 0$$

for every $n \in \mathbf{Z}$, hence, it commutes with all of $\mathfrak{d}$ and therefore belongs to the center of $\mathfrak{d}_c$ (with respect to the new bracket defined on $\mathfrak{d}_c$ and denoted by the index c), and

$$[d_m, d_n]_c = [d_m, d_n] + w(d_m, d_n) \cdot c$$

for all $m, n \in \mathbf{Z}$. It follows that $\eth_c$ is a Lie algebra with respect to the commutator $[\cdot, \cdot]_c$, when $\omega(\cdot, \cdot)$ is the above given two-cocycle. Conversely, suppose $\eth_c$ becomes a Lie algebra with respect to the commutator $[\cdot, \cdot]_c$ and some unspecified bilinear form ω, then ω can't be arbitrarily chosen.

Consider the transformation $d_0 \to b_0 = d_0$ and $d_n \to b_n = d_n - \frac{1}{n} \cdot \omega_{0,n} \cdot c$ for $n \neq 0$, where we write $\omega_{m,n}$ for $\omega(d_m, d_n)$. Dropping the c-index on the commutator on $\eth_c$, in what follows, we get

$$[b_0, b_n] = -n \cdot b_n$$

and

$$[b_m, b_n] = [d_m, d_n] = (m - n)b_{m+n} + \gamma_{m,n} \cdot c$$

where $\gamma_{m,n} = \omega_{m,n} + \frac{m-n}{m+n} \cdot \omega_{0,m+n}$ Note that $\gamma_{0,n} = 0$, for all $n \in \mathbf{Z}$. The two Lie algebras are equivalent, we have merely changed the basis as given by the transformation. Consequently, we may assume that $\omega_{0,n} = 0$ for all $n \in \mathbf{Z}$. By the Jacobi identity for d_0, d_m and d_n, we obtain

$$
\begin{aligned}
(m^2 - n^2)d_{m+n} &= (n - m)[d_0, d_{m+n}] = -[d_0, [d_m, d_n]] \\
&= [d_m, [d_n, d_0]] + [d_n, [d_0, d_m]] \\
&= n \cdot [d_m, d_n] - m \cdot [d_n, d_m] = (n + m)[d_m, d_n] \\
&= (m^2 - n^2) \cdot d_{m+n} + (m + n) \cdot \omega_{m,n} \cdot c
\end{aligned}
$$

so $(m + n) \cdot \omega_{m,n} = 0$, whence $\omega_{m,n} = \omega_m \cdot \delta_{m+n}$.

We then have the following commutation relations

$$[d_m, d_n] = (m - n) \cdot d_{m+n} + \delta_{m+n} \cdot \omega_n \cdot c$$

with $\omega_m = \omega_{m,-m}$ and $\omega_0 = 0$. The anti-symmetry of the commutator implies that $\omega_{-m} = -\omega_m$. If we use the Jacobi identity for d_k, d_m and d_n with $k + m + n = 0$, we get

$$
\begin{aligned}
0 &= [d_k, [d_m, d_n]] + [d_m, [d_n, d_k]] + [d_n, [d_k, d_m]] \\
&= ((m - n)\omega_{m+n} - (2n + m)\omega_m + (2m + n)\omega_n) \cdot c
\end{aligned}
$$

for all $m, n \in \mathbf{Z}$. Putting $n = 1$, this gives

$$(m - 1)\omega_{m+1} = (m + 2)\omega_m - (2m + 1)\omega_1$$

for every $m \in \mathbf{Z}$. This is a linear recursion relation, and since ω_1 and ω_2 determinate all ω_m, for $m \geq 3$, and then all ω_m for $m \in \mathbf{Z}$, due to the fact that $w_0 = 0$ and $w_{-m} = -w_m$, the solution space is at most two dimensional.

Observing that $w_m = m$ and $w_m = m^3$ both are solutions, it follows that the general solution is given by $w_m = \alpha \cdot m + \beta \cdot m^3$. If $\beta = 0$, we may define the transformation $d_0 \rightarrow b_0 = d_0 + \frac{1}{2}\alpha \cdot c$ and $d_n \rightarrow b_n = d_n$, for $n \neq 0$. Then $[b_0, b_n] = -nb_n$ and $[b_m, b_n] = (m-n)b_{m+n}$, which is an algebra without central charge, i.e. $c = 0$ and w_m is a coboundary. Hence the algebra for $\beta = 0$ is equivalent to $\eth$.

For a non-trivial central extension $\beta \neq 0$, we notice that the transformation $d_0 \rightarrow b_0 = d_0 + \frac{1}{2}(\alpha + \beta) \cdot c$ and $d_n \rightarrow b_n = d_n$, for $n \neq 0$, transform the commutation relations into $[b_m, b_n] = (m-n)b_{m+n} + \tilde{w}(b_m, b_n) \cdot c$ with the two-cocycle $\tilde{w}(b_m, b_n) = \delta_{m+n} \cdot \beta \cdot m(m^2 - 1)$. This transformation corresponds to a translation of the spectrum of d_0 by $\frac{1}{2}(\alpha + \beta) \cdot c$ in any representation (so it is bounded from below if d_0 is). Physically it is nothing but a translation of the energy scale and therefore having no physically significance (see below).

The factor α can therefore be chosen arbitrarily. Conventionally one selects $\alpha = -\beta$ such that $w_m = \beta(m^3 - m)$. By rescaling c, we can choose a fixed value for β. By convention $\beta = \frac{1}{12}$ is the value of choice.

Hence, we obtain the above non-trivial two-cocycle

$$w(d_m, d_n) = \delta_{m+n} \cdot \frac{m(m^2 - 1)}{12}$$

The extended Lie algebra with basis $\{c, d_n : n \in \mathbf{Z}\}$, fulfilling the commutation relations

$$[d_n, c] = 0$$

and

$$[d_m, d_n] = (m - n)d_{m+n} + \delta_{m+n} \cdot \frac{m(m^2 - 1)}{12} \cdot c$$

is the most general central extension of $\eth$, and it is called the *Virasoro algebra*, denoted *Vir*. In fact, we have proved above that every non-trivial central extension of the Lie algebra $\eth$ by a one dimensional centre is isomorphic to the Virasoro algebra.

Returning to the Sugawara construction discussed in the end of section 4.1, it follows at this point, that the operators L_n, $n \in \mathbf{Z}$ defined there, give us a representation of the Virasoro algebra with central charge given by $\frac{c \cdot \dim \mathfrak{g}}{Q+c}$, where this c is the value with regard to the considered representation of the central element giving the extended loop algebra $\hat{g}' = \tilde{g} \oplus \mathbf{C} \cdot c$.

It is also interesting to notice, that there is a natural action of $\eth$, and hence of Vir, by letting the central charge act trivially, as derivatives on the loop algebra $\tilde{g}$, which lifts to $\hat{g}'$ by acting trivially on the central element, given by

$$d_k x(n) = n \cdot x(n+k)$$

for any $x \in g$, $k, n \in \mathbf{Z}$

It is interesting to consider representations of the Virasoro algebra, or equivalently projective representations of $\eth$ with non-trivial cocycles. Let ρ_c be a representation of Vir, on a representation space V, with $\rho_c(c) = c \cdot I$, where c on the left-hand side denotes the central charge and the c on the right-hand side denotes a complex number. Then we may consider a projective representation ρ of $\eth$ with commutation relations

$$[\rho(X), \rho(Y)] = \rho([X,Y]) + \omega(X,Y) \cdot c \cdot I$$

for $X, Y \in \eth$, where the commutators involved act on the respective Lie algebras, so the new commutator appearing on the left-hand side is the common commutator of operators on V. The two representations are related by

$$\rho_c(X + \lambda \cdot c) = \rho(X) + \lambda \cdot c \cdot I$$

and

$$[X,Y]_c = [X,Y] + \omega(X,Y) \cdot c$$

Usually, one calls ρ a highest weight representation if there exists a complex number h such that the operator $D_0 = \rho(d_0)$ diagonalizes on V with eigenvalues of the form $h - n$, $n \in \mathbf{N} \setminus \{0\}$, each of the corresponding eigenspaces V_{h-n} being finite dimensional and such that $\dim V_h = 1$. The pair $(h, c) \in \mathbf{C}^2$, or sometimes just $h \in \mathbf{C}$, is called the highest weight of the highest weight representation ρ and we may choose a non-zero vector $v \in V_h$ to be named a highest weight vector.

It is well known that for each $c \in \mathbf{C}$ and $h \in \mathbf{C}$ there exists an irreducible highest weight representation $\rho = \rho_{h,c}$, on $V = V(h,c)$, which is unique up to equivalence (see [G-W 2, p. 303] or [K-R, p. 24]).

Let L_n, $n \in \mathbb{N}$, denote the basis generators of a unitary representation of the Virasoro algebra. In physics, L_0 usually denotes the energy operator and is therefore required to be bounded from below (it is selfadjoint due to the unitarity of L_0 and so its possible eigenvalues are real). If ψ is an eigenvector of L_0 with eigenvalue λ, i.e. $L_0\psi = \lambda\psi$, then $L_0 L_n\psi = (\lambda - n)L_n\psi$, for $n \in \mathbb{N}$, by use of the commutation relations for the Virasoro algebra. Hence, $L_n\psi$ is an eigenvector of L_0 with eigenvalue $\lambda - n$ or $L_n\psi = 0$. Since the spectrum of L_0 is bounded from below there exists a lowest eigenvalue $h \in \mathbb{R}$ and a corresponding eigenvector v. Then $L_0 v = hv$ and $L_n v = 0$ for all $n \in \mathbb{N}$. Thus, we arrive at highest weight representations. In physics, one seeks unitary representations, if possible, and one focuses on the irreducible ones (since such representations can be decomposed into a direct sum of irreducible ones, of which there are fewer and they are easier to handle). This is, crudely speaking, the reason for seeking irreducible unitary highest weight representations and, in fact, it is the highest weight demand which put restrictions on the possible value of the central charge and the lowest energy eigenvalue. Let L_n be such a unitary highest weight representation of the Virasoro algebra with highest weight vector v and corresponding eigenvalue h, such that $L_0 v = hv$ and $L_n v = 0$, for $n \in \mathbb{N}$. Then

$$
\begin{aligned}
\|L_{-n}v\|^2 &= \langle v, L_n L_{-n}v \rangle = \langle v, [L_n, L_{-n}]v \rangle \\
&= \left(2nh + \frac{1}{12}n(n^2 - 1)c \right) \|v\|^2
\end{aligned}
$$

wherefrom it follows that also c has to be real and, in fact, it follows that

$$
2nh + \frac{1}{12}n(n^2 - 1)c \geq 0
$$

for all $n \in \mathbb{N}$. Especially for $n = 1$ we get $h \geq 0$ and for n "large" we get $c \geq 0$. This condition is neccessary for the representation to be a unitary highest weight representation. However, it is not sufficient, since not all values of (c, h) in $[0, \infty[\times [0, \infty[$ provide us with unitary highest weight representations.

In the case $(h, c) \in \mathbb{R}^2$, $V = V(h, c)$ admits a unique (up to a scalar multiple) non-zero contravariant Hermitian form (see [G-W 2, p. 303]). It follows that if, in addition, h is non-negative and $c \geq 1$ then the highest weight representation is unitary (see [K-R, p. 26 and 27]).

We will not dwell on this fact, but refer to [K-R] and [G-W 2]. However, we will now consider the case $0 < c < 1$ briefly, using the Sugawara construction, and in the next section we will focus on the case $c = \frac{1}{2}$, by other methods. We return to the former case in section 4.3 where we treat it in all detail by use of a generalization of the Sugawara construction.

Consider any admissible representation of an arbitrary finite dimensional simple Lie algebra $\mathfrak{g}$ with basis $\{x_i\}$, $i = 1, \ldots, \dim \mathfrak{g}$ such that $\{x_i\}$, $i = 1, \ldots, \dim \mathfrak{h}$ form a basis for a subalgebra $\mathfrak{h} \subset \mathfrak{g}$ with $\mathfrak{h}$ simple.

The Lie algebra $\mathfrak{g}$ could, for example, be the extended loop algebra studied in section 4.1 (where we denoted it $\hat{\mathfrak{g}}$). Let $2Q_\mathfrak{h}$ and $2Q_\mathfrak{g}$ denote the eigenvalue of the Casimir operator of $\mathfrak{h}$, respectively $\mathfrak{g}$, in the adjoint representation of $\mathfrak{h}$, respectively $\mathfrak{g}$. We then define the Virasoro operators by the Sugawara construction for $\mathfrak{h}$, respectively $\mathfrak{g}$, i.e.

$$L_n^\mathfrak{h} = \frac{1}{Q_\mathfrak{h} + c} \sum_{i=1}^{\dim \mathfrak{h}} \sum_{k \in \mathbf{Z}} : x_i(k) x_i(n - k) :$$

and

$$L_n^\mathfrak{g} = \frac{1}{Q_\mathfrak{g} + c} \sum_{i=1}^{\dim \mathfrak{g}} \sum_{k \in \mathbf{Z}} : x_i(k) x_i(n - k) :$$

Note that that the operator c in the considered representation acts as multiplication by a scalar c different from $-Q_\mathfrak{h}$ and $-Q_\mathfrak{g}$, such that we may divide by $Q_\mathfrak{h} + c$ and $Q_\mathfrak{g} + c$, respectively. Put $L_n = L_n^\mathfrak{g} - L_n^\mathfrak{h}$, resulting in the following theorem.

Theorem 26 *The L_n's satisfy the Virasoro commutation relations with central charge $c = c_\mathfrak{g} - c_\mathfrak{h}$.*

Proof. This theorem can be found in [K-R, theorem 10.2] and [Mi, theorem 7.4.9], we follow their proofs. From the end of section 4.1 we know that

$$[L_n^\mathfrak{g}, x_i(m)] = -m \cdot x_i(m + n)$$

for $i = 1, \ldots, \dim \mathfrak{g}$ and

$$[L_n^\mathfrak{h}, x_i(m)] = -m \cdot x_i(m + n)$$

for $i = 1, \ldots, \dim \mathfrak{h}$, for all $n, m \in \mathbf{Z}$.

By subtraction we get $[L_n, x_i(m)] = 0$, for $i = 1, \ldots, \dim \mathfrak{h}$ and all $n, m \in \mathbf{Z}$. Then it follows that $[L_n, L_m^{\mathfrak{h}}] = 0$, for all $n, m \in \mathbf{Z}$, since the $L_m^{\mathfrak{h}}$'s are constructed from the $x_i(m)$, with $i = 1, \ldots, \dim \mathfrak{h}$, and

$$[L_n, L_m^{\mathfrak{h}}] = \frac{Q+c}{2} \sum_{k \in \mathbf{Z}} \sum_{i=1}^{\dim \mathfrak{h}} [L_n, x_i(k)x_i(m-k)] = 0$$

due to the fact that

$$[L_n, x_i(k)x_i(m-k)] = x_i(k)\,[L_n, x_i(m-k)] + [L_n, x_i(k)]\,x_i(m-k)$$

With $L_n^{\mathfrak{g}} = L_n^{\mathfrak{h}} - L_n$, the commutator $[L_n^{\mathfrak{g}}, L_m^{\mathfrak{g}}]$ splits into two parts

$$
\begin{aligned}
[L_n^{\mathfrak{g}}, L_m^{\mathfrak{g}}] &= [L_n^{\mathfrak{h}}, L_m^{\mathfrak{h}}] + [L_n, L_m] \\
&= (n-m)L_{n+m}^{\mathfrak{h}} + \delta_{n+m} \cdot \frac{n(n^2-1)}{12} \cdot c_{\mathfrak{h}} + [L_n, L_m]
\end{aligned}
$$

and since

$$[L_n^{\mathfrak{g}}, L_m^{\mathfrak{g}}] = (n-m)L_{n+m}^{\mathfrak{g}} + \delta_{n+m} \cdot \frac{n(n^2-1)}{12} \cdot c_{\mathfrak{g}}$$

it follows that

$$[L_n, L_m] = (n-m)L_{n+m} + \delta_{n+m} \cdot \frac{n(n^2-1)}{12} \cdot (c_{\mathfrak{g}} - c_{\mathfrak{h}})$$

implying that the L_n's satisfy the Virasoro commutation relations, with central charge $c = c_{\mathfrak{g}} - c_{\mathfrak{h}}$ $\square$

At this point we mention that the above considerations could be done for a finite dimensional semi-simple Lie algebra $\mathfrak{g} = \mathfrak{g}_1 \oplus \mathfrak{g}_2 \oplus \cdots \oplus \mathfrak{g}_q$, where for $i = 1, \ldots, q$ the $\mathfrak{g}_i$'s are simple algebras. Hence, we can construct a Virasoro algebra associated with $\mathfrak{g}$ from a highest weight representation of $\hat{\mathfrak{g}}_1 \oplus \cdots \oplus \hat{\mathfrak{g}}_q$, simply by taking the sum of the Virasoro algebras.

Define $L_n^{\mathfrak{g}}$ as

$$L_n^{\mathfrak{g}} = L_n^{\mathfrak{g}_1} + \cdots + L_n^{\mathfrak{g}_q}$$

then it follows that the terms in $L_n^{\mathfrak{g}}$ commute with each other, since the different algebras commute. Therefore the $L_n^{\mathfrak{g}}$'s fulfil the commutation relations of the Virasoro algebra, with the central charge simply being the sum of central charges (which may all be different from each other).

As an example take $\mathfrak{g}_1 = \mathfrak{g}_2 = \mathfrak{su}(2)$, the Lie algebra consisting of all 2×2 traceless anti-Hermitian matrices, corresponding to the group $Su(2)$, of unitary 2×2 matrices with determinant 1. The dimension of $\mathfrak{g}_1$ and $\mathfrak{g}_2$ is 3.

Let $\mathfrak{h}$ be the subalgebra of $\mathfrak{g} = \mathfrak{g}_1 \oplus \mathfrak{g}_2$ consisting of the diagonal, i.e. $(x, y) \in \mathfrak{h}$ if and only if $x = y$. Let $\hat{\mathfrak{g}}_1$ and $\hat{\mathfrak{g}}_2$ be the central extensions with central charge c_1 and c_2, respectively. Then the central charge of the algebra spanned by the L_n's in the theorem above is given by

$$c = \frac{3 \cdot c_1}{2 + c_1} + \frac{3 \cdot c_2}{2 + c_2} - \frac{3 \cdot (c_1 + c_2)}{2 + (c_1 + c_2)}$$

since the sum of the first two terms is the central charge of $\hat{\mathfrak{g}} = \hat{\mathfrak{g}}_1 \oplus \hat{\mathfrak{g}}_2$ and the third term is the central charge of $\hat{\mathfrak{h}}$ (we have chosen the eigenvalues of the Casimir operators to be 2, by normalization). If we had chosen c_1 and c_2 to be $c_1 = k \in \{0, 1, 2, \ldots\}$ and $c_2 = 1$ then

$$c = 1 - \frac{6}{(k + 2)(k + 3)}$$

where $k = \mathbf{N} \cup \{0\}$. The series of representations giving this series of central charges will be constructed explicitly later on by other means (see section 4.3).

From this example it appears that we have constructed a series of representations of the Virasoro algebra with central charge c_k, forming a discrete series, $0 \leq c_k < 1$. In fact, it is shown in [K-R, p. 129-138] that the irreducible representations of Vir in the highest weight representation $V(h, c)$ is unitary if and only if c is of the above form and h is of the form

$$h_k^{(r,s)} = \frac{((k + 3)r - (k + 2)s)^2 - 1}{4(k + 2)(k + 3)}$$

where $r, s \in \mathbf{N} \cup \{0\}$ such that $1 \leq s \leq r \leq k + 1$ and $k = 0, 1, 2, \ldots$ or if $c \geq 1$ and $h \geq 0$ (this formula will also be discussed later on in section 4.3). This result is very important in two-dimensional statistical mechanics models (see [B-P-Z 1]). In [B-P-Z 1, section 2] (and [B-P-Z 2, p.766-767]) the authors show that the conservation of the stress-energy tensor, of some conformally invariant two-dimensional statistical system with critical points (described by mass-less quantum field theories), gives rise to Virasoro operators (these are the coefficients appearing in the Laurent expansions of the two independent components of the stress-energy tensor), which satisfy the commutation relations of the Virasoro algebra, see also the examples in Chap. 1. Moreover, they discuss the so-called minimal theories (in section 6 and on p.769-776, respectively) and the central charge c being on the discrete form. The first few values of c_m has been confirmed by physicists in the sense that $c_1 = \frac{1}{2}$, $c_2 = \frac{7}{10}$, $c_3 = \frac{4}{5}$ and $c_4 = \frac{6}{7}$ corresponds to the Ising model, the Tri-critical Ising model, the 3-state Potts model and the Tri-critical 3-state Potts model, repectively, which are all well-known models in statistical physics.

As a another application of the above theorem we mention the so-called quantum equivalence theorem, which states that if $c = c_{\mathfrak{g}} - c_{\mathfrak{h}} = 0$ then $L_n^{\mathfrak{g}} = L_n^{\mathfrak{h}}$ for all $n \in \mathbf{Z}$ (for a proof see [Mi, p. 185]). The quantum equivalence theorem has important applications in string theory (see [Mi]).

In this section we have introduced the Virasoro algebra, shown some natural relations to the loop algebras and discussed representation theory of the Virasoro algebra. Furthermore, the presentation has been completely self-contained. Finally we emphasize the very interesting fact, both from a mathematical and a physical point of view, that all the irreducible highest weight representations of the Virasoro algebra are known ([Mi, p.174]) and that these are unitary if and only if $c \geq 1$ and $h \geq 0$ ([K-R, p.88-89]) or $c = c_k = 1 - \frac{6}{(k+2)(k+3)}$, and $h \in \{h_k^{(r,s)} : r, s \subset \mathbf{N} \cup \{0\}, 1 \leq s \leq r \leq k+1\}$, for $k \in \mathbf{N} \cup \{0\}$ ([K-R, p.129-138]).

In the following section we will consider some special representations of the Virasoro algebra, with central charge $c = \frac{1}{2}$. To our knowledge the constructions have not been discussed in detail before; however, it is proposed by Kac and Raina in [K-R]. In section 4.3 we construct a series of representations with central charge c running through the above series, all c in $[0, 1[$.

4.3 Representation of *Vir* with $c = \frac{1}{2}$

In this section we will construct some explicit representations of the Virasoro algebra *Vir*, with central charge $c = \frac{1}{2}$. As representation space we will use the anti-symmetric Fock Hilbert space $\mathcal{F}_\wedge(\mathcal{H})$, as discussed in chapter 2.

To our knowledge these constructions have not been considered in detail earlier. As some basis material for the idea of this section we mention [P-S, section 9.2], [K-R, chapter 2 and 3 together with corollary 12.1], [G-W 2, section 1] and [Jø, definition 8.5.1] beyond chapter 2 and sections 4.1 and 4.2 in this book.

Consider the so-called *fermionic oscillator algebra* generated by

$$\{x_n : n \in \mathbf{Z}_q\}$$

where $\mathbf{Z}_q$ denotes the set $\mathbf{Z} + q$, i.e. $n \in \mathbf{Z}_q$ if and only if $n - q \in \mathbf{Z}$, fulfilling the anti-commutation relations

$$[x_m, x_n]_+ = x_m x_n + x_n x_m = \delta_{m+n}$$

for $m, n \in \mathbf{Z}_q$. We will only consider $q = 0$ and $q = \frac{1}{2}$, giving the two essentially different cases. In the case $q = 0$ we talk about the *Ramond sector* and in the case $q = \frac{1}{2}$ we talk about the *Neveu-Schwarz sector*. The essential difference between the two cases is that in the Ramond sector, $q = 0$, the index $m = 0 \in \mathbf{Z}_0$ is allowed (as is the case for arbitrary $q \in \mathbf{Z}$), but in the Neveu-Schwarz sector, $q = \frac{1}{2}$, the index $m = 0 \notin \mathbf{Z}_q$ is not allowed (as is the case for arbitrary $q \notin \mathbf{Z}$). This is essential since an element x_0, which appears only in the Ramond sector, is a distinguished element because of the anti-commutator relations. In fact, $2x_0 = [x_0, x_0]_+ = 1$ implies $x_0 = \frac{1}{2}$.

In the following, we construct a representation in the fermionic Fock Hilbert space $\mathcal{F}_\wedge(\mathcal{H})$, modelled over a given Hilbert space $\mathcal{H}$, and thereby construct a representation of the Virasoro algebra with central charge $c = \frac{1}{2}$.

Let $\{e_n\}_{n \in \mathbf{N}_q}$ denote two orthonormal bases in the Hilbert space $\mathcal{H}$, where $\mathbf{N}_q = \mathbf{N}_0 = \mathbf{N} \cup \{0\}$ for $q = 0$ and $\mathbf{N}_q = \mathbf{N}_{\frac{1}{2}} = \mathbf{N}_0 + \frac{1}{2}$ for $q = \frac{1}{2}$, i.e. $\mathbf{N}_q = \mathbf{N}_0 + q$. We define the representations (linear homomorphisms of the algebra into an algebra of linear operators) π_q by $\pi_q(x_n) = a(e_n)$ for $n \in \mathbf{Z}_q$ positive and $\pi_q(x_n) = a^*(e_{-n})$ for $n \in \mathbf{Z}_q$ negative. Here $a(\cdot)$ and $a^*(\cdot)$ denote, repectively, the annihilation and the creation operators in the Fock representation of the CAR-algebra, discussed in section 2.2 of chapter 2.

In the Ramond sector $q = 0$, we additionally define

$$\pi_0(x_0) = \frac{1}{\sqrt{2}} \left(a(e_0) + a^*(e_0) \right)$$

For notational reasons we introduce the abbreviation $a_n = a(e_n)$ and $a_n^* = a^*(e_n)$, where $n \in \mathbf{N}_q$. By the anti-commutation relations of the CAR-algebra, $[a_m, a_n^*]_+ = \langle e_m, e_n \rangle \cdot I = \delta_{m-n} \cdot I$ and $[a_m, a_n]_+ = 0 = [a_m^*, a_n^*]_+$. It follows that the mappings indeed define homomorphisms and therefore representations of the fermionic oscillator algebras, conserving the anti-commutator relations, as intended.

We may define an anti-linear anti-involution α on the fermionic oscillator algebras by $\alpha(x_n) = x_{-n}$, for $n \in \mathbf{Z}_q$. It is consistent with the involution on the CAR-algebra,

$$\pi_q(x_n)^* = a_n^* = \pi_q(x_{-n}) = \pi_q(\alpha(x_n))$$

for $n \in \mathbf{Z}_q$ positive, and

$$\pi_q(x_n)^* = (a_{-n}^*)^* = a_{-n} = \pi_q(x_{-n}) = \pi_q(\alpha(x_n))$$

for $n \in \mathbf{Z}_q$ negative, and in the case of the Ramond sector

$$\pi_0(x_0)^* = \frac{1}{\sqrt{2}}(a_0 + a_0^*) = \pi_0(x_0) = \pi_0(\alpha(x_0))$$

Let $\langle \cdot, \cdot \rangle$ denote the inner product on $\mathcal{F}_\wedge(\mathcal{H})$. Since the inner product is a positive definite contravariant Hermitian form with respect to the Hilbert space adjoint, as the anti-linear anti-involution, the representations become unitary.

In the following we discard the representation symbols π_q and simply write x_k for $\pi_q(x_k)$, hence we identify a_k with x_k, a_k^* with x_{-k} for $k \in \mathbf{Z}_q$ positive and in the case of the Ramond sector $\frac{1}{\sqrt{2}}(a_0 + a_0^*)$ with x_0.

From another point of view, we merely rename the operators, calling a_k for x_k, a_k^* for x_{-k}, for $k \in \mathbf{Z}_q$ positive and in the Ramond sector $\frac{1}{\sqrt{2}}(a_0 + a_0^*)$ for x_0. Notice that we have $x_k^* = x_{-k}$. Thus, we are considering (a modification of) the Fock representation of the CAR-algebra and could have started our construction at this point. However, the above discussion provides a better understanding of the construction, in view of our treatment in section 4.2 and the analogous construction for the bosonic oscillator or Heisenberg algebra (see for example [K-R, p. 12]).

In accordance with the anti-symmetry in $\mathcal{F}_\wedge(\mathcal{H})$ we define normal ordering $:x_j x_k:$ of $x_j x_k$ by

$$:x_j x_k: = \begin{cases} x_j x_k & \text{for } j \leq k \\ -x_k x_j & \text{for } j > k \end{cases}$$

for $j, k \in \mathbf{Z}_q$. Notice that this defines bounded operators on $\mathcal{F}_\wedge(\mathcal{H})$, since both the annihilation and the creation operators are bounded on $\mathcal{F}_\wedge(\mathcal{H})$. Now, for any $k \in \mathbf{Z}$ we define the Virasoro operators on $\mathcal{F}_\wedge(\mathcal{H})$ by

$$L_k = \frac{1}{2} \sum_{j \in \mathbf{Z}_q} j : x_{-j} x_{j+k} : + \delta_k \cdot \frac{1 - 2 \cdot q}{16}$$

where the last term may also be written as $\delta_k \cdot \delta_q \cdot \frac{1}{16}$ and $q = 0, \frac{1}{2}$ as usual. Notice that indices on the L_k's are always integers independent of q. It turns out that these Virasoro operators are unbounded on $\mathcal{F}_\wedge(\mathcal{H})$, which is why we have to specify their domains.

The product basis vectors $e_{i_1} \wedge \ldots \wedge e_{i_n}$, $i_1 < \cdots < i_n$, $n \in \mathbf{N}$ (see section 2.1), form together with Ω a basis of $\mathcal{F}_\wedge(\mathcal{H})$. Collect the basis product vectors with the same index sum in $\mathbf{M}_q$, where $\mathbf{M}_q$ is $\mathbf{N}_0$ for $q = 0$ and $\mathbf{N}_{\frac{1}{2}} \cup \mathbf{N}_0 \setminus \{1\}$ for $q = \frac{1}{2}$, i.e. for a given $m \in \mathbf{M}_q \setminus \{0\}$ let

$$B_m = \operatorname{span} \left\{ e_{i_1} \wedge \cdots \wedge e_{i_n} : \sum_{k=1}^{n} i_k = m, \ n \in \mathbf{N} \right\}$$

and for $m = 0$ let $B_0 = \operatorname{span}\{\Omega\}$ for $q = \frac{1}{2}$, and $B_0 = \operatorname{span}\{\Omega, e_0\}$ for $q = 0$.

Observe that the B_m, $m \in \mathbf{M}_q$, are mutually orthogonal and span all of $\mathcal{F}_\wedge(\mathcal{H})$, moreover, each B_m is finite dimensional by construction. So we have a *grading* of $\mathcal{F}_\wedge(\mathcal{H})$, as the Hilbert space direct sum of all B_m, $m \in \mathbf{M}_q$, i.e.

$$\mathcal{F}_\wedge(\mathcal{H}) = \oplus_{m \in \mathbf{M}_q} B_m$$

Moreover, we define $\mathcal{D}_0$ as the algebraic direct sum of the B_m, $m \in \mathbf{M}_q$, i.e.

$$\mathcal{D}_0 = \oplus_{alg} B_m$$

Then $\mathcal{D}_0$ is evidently a dense subspace of $\mathcal{F}_\wedge(\mathcal{H})$, which we call *the finite energy subspace* of $\mathcal{F}_\wedge(\mathcal{H})$, since it turns out below that vectors in $\mathcal{D}_0$ represent particles in $\mathcal{F}_\wedge(\mathcal{H})$ with finite energy.

We choose $\mathcal{D}_0$ as the common domain for all the L_k, $k \in \mathbf{Z}$, though it is obvious that $\mathcal{D}_0 = \mathcal{D}_0(q)$ depends on q in the same way that $B_m = B_m(q)$ does. The reason for this choice of domain is that all the infinite sums appearing in the definition of the L_k's reduce to finite sums on $\mathcal{D}_0$. Notice, in this context, that the normal ordering in the definition of L_k does not, in general, change the product for $k \neq 0$. In fact $: x_{-j}x_{j+k} := x_{-j}x_{j+k}$, for $j \geq -\frac{k}{2}$, and $: x_{-j}x_{j+k} := -x_{j+k}x_{-j} = x_{-j}x_{j+k} - \delta_k$ for $j < -\frac{k}{2}$, where we have used the anti-commutator relations. So, for $k \in \mathbf{Z} \setminus \{0\}$

$$L_k = \frac{1}{2} \sum_{j \in \mathbf{Z}_q} j \cdot x_{-j}x_{j+k}$$

However, for $k = 0$ the normal ordering contributes formally with an infinite sum, but on $\mathcal{D}_0$ it reduces to a finite sum.

For any F in $\mathcal{D}_0$, there exists a $M \in \mathbf{M}_q$ such that $F_m = 0$, for $m > M$, where $F = \oplus_{m \in \mathbf{M}_q} F_m$, with each $F_m \in B_m$. Then, for any $k \in \mathbf{Z}$

$$
\begin{aligned}
L_k F &= \frac{1}{2} \left(\sum_{j \geq -\frac{k}{2}} j \cdot x_{-j}x_{j+k} - \sum_{j < -\frac{k}{2}} j \cdot x_{j+k}x_{-j} \right) F \\
&= \frac{1}{2} \sum_{-\frac{k}{2} \leq j \leq M-k} j \cdot x_{-j}x_{j+k}F - \frac{1}{2} \sum_{-M \leq j < -\frac{k}{2}} j \cdot x_{j+k}x_{-j}F
\end{aligned}
$$

since each F_m in $F = \oplus_{m \in \mathbf{M}_q} F_m$ may be written as a limit of linear combinations of the basis product vector $e_{i_1} \wedge \ldots \wedge e_{i_n}$, with $\sum_{l=1}^{n} i_l = m \in \mathbf{M}_q$ consisting of (one-particle) basis vectors e_{i_l} with indices $i_l \in \mathbf{N}_q$ less that or equal to m.

In the previous paragraphs we have used that $x_{j+k} = a(e_{j+k})$ annihilates, for $j > M - k$, and vanishes when applied to F_m, and that $x_{-j} = a(e_{-j})$ annihilates, for $j < -M$, and also vanishes when applied to F_m. Observe that $L_k : B_m \rightarrow B_{m+k}$ and leaves $\mathcal{D}_0$ invariant. Thus, we have shown that the formally infinite sums in the definition of the L_k, $k \in \mathbf{Z}$, reduce to finite sums on the dense invariant domain $\mathcal{D}_0$. In the special case $k = 0$ we have

$$
\begin{aligned}
L_0 &= \frac{1}{2} \sum_{j \in \mathbf{N}_q} j \cdot x_{-j}x_j - \frac{1}{2} \sum_{-j \in \mathbf{N}_q} j \cdot x_j x_{-j} + \frac{1 - 2q}{16} \\
&= \sum_{j \in \mathbf{N}_q} j \cdot x_{-j}x_j + \frac{1 - 2q}{16}
\end{aligned}
$$

We will call L_0 *the energy operator*, since the eigenvalues for L_0 are given by $h_q + m$, with corresponding eigenspaces B_m, where $h_q = \frac{1-2q}{16}$. Observe that L_0 is unbounded, but bounded from below by h_q, since its spectrum is bounded from below. Further, notice that $\mathcal{D}_0$ consists of physical states with finite energy, hence the name.

Our aim is to construct representations of the Virasoro algebra. If the L_k's fulfil the commutation relations of the Virasoro algebra, then we have some *positive energy representations* ($h_q \geq 0$) which are also *finite energy representations* on $\mathcal{D}_0$ (for general definitions see [P-S, p.171]).

Below we show that L_k, $k \in \mathbf{Z}$, indeed fulfil the commutation relations of the Virasoro algebra.

Lemma 27 *The operators L_k, $k \in \mathbf{Z}$, fulfil*

$$[x_m, L_k] = \left(m + \frac{1}{2}k\right) x_{m+k}$$

on $\mathcal{D}_0$, for all $m \in \mathbf{Z}_q$.

Proof. From above it follows that all formally infinite sums appearing below reduce to finite sums on $\mathcal{D}_0$, and that the normal ordering in the definitions of the L_k, $k \in \mathbf{Z}$ only contributes for $k = 0$. For $k \in \mathbf{Z} \setminus \{0\}$ we have

$$
\begin{aligned}
[x_m, L_k] &= \frac{1}{2} \sum_{j \in \mathbf{Z}_q} j \cdot [x_m, x_{-j} x_{j+k}] \\
&= \frac{1}{2} \sum_{j \in \mathbf{Z}_q} j \left([x_m, x_{-j}]_+ x_{j+k} - x_{-j} [x_m, x_{j+k}]_+\right) \\
&= \frac{1}{2} \sum_{j \in \mathbf{Z}_q} j \left(\delta_{m-j} \cdot x_{j+k} - x_{-j} \cdot \delta_{j+m+k}\right) \\
&= \frac{1}{2} (m \cdot x_{m+k} + (m + k) \cdot x_{m+k}) \\
&= \left(m + \frac{1}{2}k\right) \cdot x_{m+k}
\end{aligned}
$$

where we have used the anti-commutator relations and the formula

$$[x_m, x_n x_l] = [x_m, x_n]_+ x_l - x_n [x_m, x_l]_+$$

for all allowed indices m, n and l. For $k = 0$ we get

$$[x_m, L_0] = \sum_{j \in \mathbf{N}_q} j \left([x_m, x_{-j}]_+ x_j - x_{-j} [x_m, x_j]_+\right)$$

$$= \sum_{j \in \mathbf{N}_q} (j \cdot \delta_{m-j} \cdot x_j - j \cdot x_{-j} \cdot \delta_{j+m})$$

$$= \sum_{j \in \mathbf{Z}_q} j \cdot x_j \cdot \delta_{m-j}$$

$$= m \cdot x_m$$

where we, moreover, have used that the scalar term (times the unit operator) appearing in L_0 doesn't influence the commutator.   □

We shall use the above lemma to prove the proposition below.

Proposition 28 *The* L_m, $m \in \mathbf{Z}$, *fulfil the commutator relations of the Virasoro algebra with central charge* $c = \frac{1}{2}$, *i.e.*

$$[L_m, L_n] = (m - n) \cdot L_0 + \delta_{m+n} \cdot \frac{m^3 - m}{24}$$

for all $m, n \in \mathbf{Z}$.

Proof. As in lemma 27 all sums appearing below reduce to finite sums on $\mathcal{D}_0$. In the case of $q = 0$, all the indices on the e_j's will belong to $\mathbf{N}_0$ and all other indices, i.e. the indices on the x_j's and on the L_k's, belong to $\mathbf{Z}$. In the case $q = \frac{1}{2}$ all the indices on the e_j's will belong to $\mathbf{N}_{\frac{1}{2}} = \mathbf{N} - \frac{1}{2}$ and the indices on the x_j's will belong to $\mathbf{Z}_{\frac{1}{2}} = \mathbf{Z} - \frac{1}{2}$, but the indices on the L_k's will belong to $\mathbf{Z}$.

Remember that the normal ordering in $L_m = \frac{1}{2}\sum_{j \in \mathbf{Z}} j :x_{-j}x_{j+m}:$ doesn't contribute unless $m = 0$, since $:x_j x_{j+m}: = x_{-j}x_{j\,|\,m} - \delta_m$.

Since it is trivial that $[L_0, L_0] = 0$, we can treat the case of $m = 0$, for $n \in \mathbf{Z} \setminus \{0\}$, by $[L_0, L_n] = -[L_n, L_0]$. This implies that the case for $m = 0$ follows directly from the general case, since our treatment will not concern the normal ordering appearing in L_n.

For arbitrary $m \in \mathbf{Z} \setminus \{0\}$ and $n \in \mathbf{Z}$ we have

$$[L_m, L_n] = \frac{1}{2}\sum_{j \in \mathbf{Z}_q} j \cdot (x_{-j}[x_{j+m}, L_n] + [x_{-j}, L_n]x_{j+m})$$

$$= \frac{1}{2}\sum_{j \in \mathbf{Z}_q} j \cdot \left(j + m + \frac{1}{2}n\right) x_{-j}x_{j+m+n}$$

$$+ \frac{1}{2}\sum_{j \in \mathbf{Z}_q} j \cdot \left(-j + \frac{1}{2}n\right) x_{-j+n}x_{j+n}$$

Let $-j+n=-i$, such that $j=i+n$ and $j+m=i+m+n$, and write j instead of i, in the second sum, i.e. we make the transformation $j \to j+n$, then

$$[L_m, L_n]$$

$$= \frac{1}{2} \sum_{j \in \mathbf{Z}_q} \left(j \cdot \left(j+m+\frac{1}{2}n \right) - (j+n)\left(j+\frac{1}{2}n \right) \right) x_{-j} x_{j+m+n}$$

$$= \frac{1}{2}(m-n) \sum_{j \in \mathbf{Z}_q} j \cdot x_{-j} x_{j+m+n} - \frac{n^2}{4} \sum_{j \in \mathbf{Z}_q} x_{-j} x_{j+m+n}$$

$$= (m-n) \cdot L_{m+n} - (m-n)\frac{1}{2} \sum_{j \in I_q} j \cdot \delta_{m+n} - \frac{n^2}{4} \sum_{j \in \mathbf{Z}_q} x_{-j} x_{j+m+n}$$

where in the last equality we have used $: x_{-j} x_{j+m+n} := x_{-j} x_{j+m+n}$ for $j \geq -\frac{m+n}{2}$ and that $: x_{-j} x_{j+m+n} := -x_{j+m+n} x_{-j} = x_{-j} x_{j+m+n} + \delta_{m+n}$ for $j < -\frac{m+n}{2}$.

Notice that we are only summing over a finite set I_q in the second term, since only finitely many terms in $\sum_{j \in \mathbf{Z}_q} j \cdot x_{-j} x_{j+m+n}$ are non-vanishing on $\mathcal{D}_0$. We have

$$-(m-n)\frac{1}{2} \sum_{j \in I_q} j \cdot \delta_{m+n} = -\delta_{m+n} \cdot m \cdot \sum_{j \in I_q} j$$

Moreover, the last sum also reduces to a finite sum. In fact, using

$$x_{-j} x_{j+m+n} + x_{j+m+n} x_{-j} = \delta_{m+n}$$

and

$$\sum_{j \in \mathbf{Z}_q} x_{-j} x_{j+m+n} = \sum_{j \in \mathbf{Z}_q} x_{j+m+n} x_{-j}$$

due to the transformation $j \to -(j+m+n)$, we obtain

$$\sum_{j \in \mathbf{Z}_q} x_{-j} x_{j+m+n} = \frac{1}{2} \sum_{j \in \mathbf{Z}_q} (x_{-j} x_{j+m+n} + x_{j+m+n} x_{-j})$$

$$= \frac{1}{2} \sum_{j \in J_q} \delta_{m+n} = \delta_{m+n} \cdot \frac{1}{2} \text{card}(J_q)$$

where J_q is a finite subset of $\mathbf{Z}_q$, since $\sum_{j \in \mathbf{Z}_q} x_{-j} x_{j+m+n}$ reduces to a finite sum on $\mathcal{D}_0$. Thus, the cardinality $\text{card}(J_q)$ of J_q is finite. Hence, for $m, n \in \mathbf{Z}$ we have

$$[L_m, L_n] = (m-n)L_{m+n} + \delta_{m+n} \cdot c_m(q)$$

where $c_m(q) \in \mathbf{C}$ is determined by

$$c_m(q) = \langle \Omega, ([L_m, L_{-m}] - 2mL_0) \, \Omega \rangle$$

It follows immediately that $c_0(q) = 0$ and $c_{-m}(q) = -c_m(q)$. Therefore we only need to calculate $c_m(q)$ for $m \in \mathbf{N}$.

Observe that

$$L_0 \cdot \Omega = \sum_{j \in \mathbf{N}_q} j \cdot x_{-j} x_j \Omega + \frac{1 - 2q}{16} \cdot \Omega = \frac{1 - 2q}{16} \cdot \Omega$$

since $j \cdot x_j \Omega = 0$ for $j \geq 0$. Moreover, for $m \in \mathbf{N}$ we have

$$L_m \Omega = \frac{1}{2} \left(\sum_{j \geq -\frac{m}{2}} j \cdot x_{-j} x_{j+m} - \sum_{j < -\frac{m}{2}} j \cdot x_{j+m} x_{-j} \right) \Omega = 0$$

since $x_{j+m} \Omega = 0$ for $j \geq -\frac{m}{2}$ and $x_{-j} \Omega = 0$ for $j < -\frac{m}{2}$. It follows that

$$c_m(q) = \|L_{-m} \Omega\|^2 - (1 - 2q) \cdot \frac{m}{8}$$

for $m \in \mathbf{N}$, where we have used that $L_m^* = L_{-m}$, which is a consequence of

$$\left\langle G, \sum_{j \geq -\frac{m}{2}} x_{-j} x_{j+m} F \right\rangle = \left\langle \sum_{j \geq -\frac{m}{2}} x_{-j-m} x_j G, F \right\rangle$$

$$= \left\langle \sum_{j \geq \frac{m}{2}} x_{-j} x_{j-m} G, F \right\rangle$$

and similarly

$$\left\langle G, \sum_{j < -\frac{m}{2}} x_{j+m} x_{-j} F \right\rangle = \left\langle \sum_{j < \frac{m}{2}} x_{j+m} x_{-j} G, F \right\rangle$$

for all $G, F \in \mathcal{D}_0$.

We now calculate $L_{-m} \Omega$ for $m \in \mathbf{N}$. We have

$$L_{-m} \Omega = \frac{1}{2} \left(\sum_{j \geq \frac{m}{2}} j \cdot x_{-j} x_{j-m} - \sum_{j < \frac{m}{2}} j \cdot x_{j-m} x_{-j} \right) \Omega$$

Since $x_{j-m}\Omega = 0$ for $j > m$ and $x_{-j}\Omega = 0$ for $j < 0$, the sums become finite as desired and

$$L_{-m}\Omega = \frac{1}{2}\left(\sum_{\frac{m}{2} \leq j \leq m} j \cdot x_{-j}x_{j-m} - \sum_{0 \leq j < \frac{m}{2}} j \cdot x_{j-m}x_{-j} \right)\Omega$$

Transform j into $m-j$ ($\in \mathbf{Z}_q$) in the first sum, such that $-j \to j-m$ and $j - m \to -j$, and the summation over j, $\frac{m}{2} \leq j \leq m$, becomes a summation over j where $0 \leq j \leq \frac{m}{2}$. Hence, we get

$$L_{-m}\Omega = \frac{1}{2} \sum_{0 \leq j < \frac{m}{2}} (m - 2j)x_{j-m}x_{-j}\Omega \tag{4.1}$$

In the case of m even, we should add a term $\frac{1}{2}x_{-\frac{m}{2}}x_{-\frac{m}{2}}\Omega = \frac{1}{2}e_{\frac{m}{2}} \wedge e_{\frac{m}{2}} = 0$, however, it vanishes and is disposed. For further treatment we have to separate equation (4.1) into the two distinct cases of the Ramond sector ($q = 0$) and the Neveu-Schwarz sector ($q = \frac{1}{2}$) and, moreover, subdivide these cases into the distinct subcases of m even and m odd.

Consider first the case of the Ramond sector, $q = 0$. Then

$$L_{-m}\Omega = \frac{1}{2} \sum_{0 \leq j < \frac{m}{2}} (m - 2j)\tilde{e}_{m-j} \wedge \tilde{e}_j$$

where $\tilde{e}_0 = \frac{1}{\sqrt{2}}e_0$ and $\tilde{e}_j = e_j$ for $0 < j < \frac{m}{2}$. The largest value of j in the sum is different dependent on the parity of m. For m even, $\frac{m}{2} - 1$ is the largest value of j in the sum, and for m odd, $\frac{m-1}{2}$ is the largest value of j in the sum. First, we will consider the subcase of m even. Observing that the product vectors appearing in the sum are normalized and form an orthogonal system with $\|\tilde{e}_m \wedge \tilde{e}_0\|^2 = \frac{1}{2}$. We obtain

$$\|L_{-m}\Omega\|^2 = \frac{1}{4} \sum_{j=1}^{\frac{m}{2}-1}(m - 2j)^2 + \frac{1}{4} \cdot \frac{m^2}{2}$$

$$= \frac{1}{4} \sum_{j=1}^{\frac{m}{2}-1}(2j)^2 + \frac{m^2}{8}$$

$$= \sum_{j=1}^{\frac{m}{2}-1}j^2 + \frac{m^2}{8}$$

where we have used the transformation $j \to -j + \frac{m}{2}$. Now for $m \in \mathbf{N}$ even

$$\sum_{j=1}^{\frac{m}{2}-1} j^2 = \frac{m^3 - 3m^2 + 2m}{24}$$

by induction. So

$$\|L_{-m}\Omega\|^2 = \frac{m^3 - 3m^2 + 2m}{24} + \frac{m^2}{8} = \frac{m^3 + 2m}{24}$$

for $m \in \mathbf{N}$ even. Hence

$$c_m(0) = \|L_{-m}\Omega\|^2 - \frac{m}{8} = \frac{m^3 - m}{24}$$

for $m \in \mathbf{N}$ even.

Consider now the subcase of m odd. As above the product vectors, in the sum of $L_{-m}\Omega$, are normalized and form an orthogonal system, except for $j = 0$, where $\|\tilde{e}_m \wedge \tilde{e}_0\|^2 = \frac{1}{2}$. Therefore we get

$$\|L_{-m}\Omega\|^2 = \frac{1}{4} \sum_{j=1}^{\frac{m-1}{2}} (m - 2j)^2 + \frac{1}{4} \cdot \frac{m^2}{2}$$

$$= \frac{1}{4} \sum_{j=1}^{\frac{m-1}{2}} (2j - 1)^2 + \frac{m^2}{8}$$

due to the transformation $j \to -j + \frac{m+1}{2}$. Now, for $m \in \mathbf{N}$ odd

$$\sum_{j=1}^{\frac{m-1}{2}} (2j - 1)^2 = \frac{m^3 - 3m^2 + 2m}{6}$$

by induction. So

$$\|L_{-m}\Omega\|^2 = \frac{1}{4} \cdot \frac{m^3 - 3m^2 + 2m}{6} + \frac{m^2}{8} = \frac{m^3 + 2m}{24}$$

Hence, for $m \in \mathbf{N}$ odd

$$c_m(0) = \|L_{-m}\Omega\|^2 - \frac{m}{8} = \frac{m^3 - m}{24}$$

Notice that the final formula for $c_m(0)$ is independent of the parity of m. In the Ramond sector $(q = 0)$, we have

$$c_m(0) = \frac{m^3 - m}{24}$$

for all $m \in \mathbf{N}$, hence for all $m \in \mathbf{Z}$, since $c_0(0) = 0$ and $c_{-m}(0) = -c_m(0)$.

We now consider the Neveu-Schwarz sector, where $q = \frac{1}{2}$, which we treat in a similar manner. Here we have

$$L_{-m}\Omega = \frac{1}{2} \sum_{0 < j < \frac{m}{2}} (m - 2j)e_{m-j} \wedge e_j$$

since $j \in \mathbf{Z}_{\frac{1}{2}}$. Observing that the product vectors appearing in the sum form an orthonormal set. We obtain

$$\|L_{-m}\Omega\|^2 = \frac{1}{4} \sum_{0 < j < \frac{m}{2}} (m - 2j)^2$$

As in the case $q = 0$ we split the further treatment of the case $q = \frac{1}{2}$ into two separate subcases – one for m even and one for m odd, since the largest value of j in the sum depends on the parity of $m \in \mathbf{N}$.

Consider first the subcase of $m \in \mathbf{N}$ even. We get

$$\|L_{-m}\Omega\|^2 = \frac{1}{4} \sum_{j=\frac{1}{2}}^{\frac{m-1}{2}} (m - 2j)^2 = \frac{1}{4} \sum_{k=1}^{\frac{m}{2}} (2k - 1)^2$$

where $j \in \mathbf{N}_{\frac{1}{2}}$, but $k = \frac{m}{2} - \left(j - \frac{1}{2}\right) \in \mathbf{N}$. It is well-known that

$$\sum_{k=1}^{\frac{m}{2}} (2k - 1)^2 = \frac{1}{6}(m^3 - m)$$

by induction. Hence

$$c_m\left(\frac{1}{2}\right) = \|L_{-m}\Omega\|^2 = \frac{1}{4} \cdot \frac{1}{6}(m^3 - m) = \frac{m^3 - m}{24}$$

for $m \in \mathbf{N}$ even.

Consider now the subcase where m is odd. We have

$$\|L_{-m}\Omega\|^2 = \frac{1}{4}\sum_{j=\frac{1}{2}}^{\frac{m}{2}-1}(m-2j)^2 = \frac{1}{4}\sum_{k=1}^{\frac{m-1}{2}}(2k)^2 = \sum_{k=1}^{\frac{m-1}{2}}k^2$$

where $j \in \mathbf{N}_{\frac{1}{2}}$, but $k = \frac{m-1}{2} - \left(j - \frac{1}{2}\right) \in \mathbf{N}$. It easily follows that

$$\sum_{k=1}^{\frac{m-1}{2}}k^2 = \frac{m^3-m}{24}$$

by induction. Hence

$$c_m\left(\tfrac{1}{2}\right) = \|L_{-m}\Omega\|^2 = \frac{m^3-m}{24}$$

for $m \in \mathbf{N}$ odd.

As in the case of $q = 0$ it turns out that the final formula for $c_m\left(\frac{1}{2}\right)$ is independent of the parity of m. In the Neveu-Schwarz sector we have

$$c_m\left(\frac{1}{2}\right) = \frac{m^3-m}{24}$$

for all $m \in \mathbf{N}$, hence for all $m \in \mathbf{Z}$, since $c_0\left(\frac{1}{2}\right) = 0$ and $c_{-m}\left(\frac{1}{2}\right) = -c_m\left(\frac{1}{2}\right)$. Observe furthermore that $c_m(0) = c_m\left(\frac{1}{2}\right) = c_m$ is independent of the sector considered, $q = 0, \frac{1}{2}$. That is, in both sectors we end up with the commutation relations

$$[L_m, L_n] = (m-n)L_{m+n} + \delta_{n+n} \cdot \frac{m^3-m}{24}$$

for all $m, n \in \mathbf{Z}$. Hereby the proposition is proved. $\qquad\qquad\square$

Hence, we have constructed some positive and finite energy representations ρ_q of the Virasoro algebra $Vir = \oplus_{n\in\mathbf{Z}}\mathbf{C}\cdot d_n \oplus \mathbf{C}\cdot c$ in the anti-symmetric Fock Hilbert space, by

$$\rho_q(d_n) = L_n$$

where $q = 0, \frac{1}{2}$. However, it turns out that these representations of Vir are reducible. In fact, the orthogonal subspaces $\mathcal{F}_{\wedge}^+(\mathcal{H}) = \oplus_{n=0}^{\infty}(\wedge^{2n}\mathcal{H})$ and $\mathcal{F}_{\wedge}^-(\mathcal{H}) = \oplus_{n=0}^{\infty}(\wedge^{2n+1}\mathcal{H})$, consisting of an even and an odd number of particles, respectively, are both invariant due to

the fact that all L_k, $k \in \mathbf{Z}$, either create or annihilate two particles or conserve the number of particles. Hence, each L_k, $k \in \mathbf{Z}$,
conserves the parity of the number of particles in a given physical
state. It turns out that the subrepresentations $\rho_q^\pm$ of the Virasoro
algebra in $\mathcal{F}_\wedge^\pm(\mathcal{H})$ are irreducible. We return to this question below.
Naturally the subrepresentations are still positive and finite energy
representations, with the common domain

$$\mathcal{D}_0 \cap \mathcal{F}_\wedge^\pm(\mathcal{H}) = \mathcal{D}_0^\pm$$

for all the unbounded Virasoro operators $L_k, k \in \mathbf{Z}$.

We now modify the previously given definition of highest weight
representations of Vir (following [G-W 2, p. 303] and [Jø, p. 238])
to lowest weight representations, trying to include our interesting
cases. We call a representation ρ of $Vir = \oplus_{n \in \mathbf{Z}} \mathbf{C} \cdot d_n \oplus \mathbf{C} \cdot c$, with
a dense domain $\mathcal{D}_\rho = \cap_{n \in \mathbf{Z}} \mathcal{D}(\rho(d_n))$ in a Hilbert space $\mathcal{K}$, a *lowest
weight representation* if;

1. the operator $\rho(d_0)$ diagonalizes on $\mathcal{D}_\rho$ with eigenvalues of the
   form $h + n$, $n \in \mathbf{N}_0$ (except for a few $n \in \mathbf{N}$), and each of the
   corresponding eigenspaces V_{h+n} is finite dimensional.

2. especially $\dim V_h = 1$.

3. $\rho(c) = c \cdot I$, where c denotes both the central element and a
   complex number.

We note that the bracket in condition 1 is a modification relative
to [G-W 2] and [Jø]. This modification is necessary in our case (see
below). Usually the pair (h, c), and sometimes just h, is called the
lowest weight and a non-zero normalized vector F_0 in V_h is called
a lowest weight vector. The lowest weight vector is not necessarily
cyclic. Observe that if ρ is unitary, as in our cases, since

$$\langle F, L_m G \rangle = \langle L_{-m} F, G \rangle$$

for all $m \in \mathbf{Z}$ and all $F, G \in \mathcal{D}_0$, then the lowest weight h is real,
$h = \langle F_0, L_0 F_0 \rangle = \langle L_0 F_0, F_0 \rangle = \bar{h}$.

Consider first ρ_q^+, where $\mathcal{K} = \mathcal{F}_\wedge^+(\mathcal{H})$ and $\mathcal{D}_\rho = \mathcal{D}_0^+$. We have
$L_m = \rho_q^+(d_m)$, $m \in \mathbf{Z}$. We have shown that L_0 diagonalizes on $\mathcal{D}_0^+$,
with eigenvalues of the form $h_q + m$.

If $q = 0$ then $h_0 = \frac{1}{16}$, $L_0 \Omega = h_0 \Omega$ with $\Omega \in \mathcal{F}_\wedge^+(\mathcal{H})$ and $m \in$
$\mathbf{N}_0 \setminus \{1, 2\}$, since $e_1 \notin \mathcal{F}_\wedge^+(\mathcal{H})$ and $e_1 \wedge e_1 = 0$.

If $q = \frac{1}{2}$ then $h_{\frac{1}{2}} = 0$, again $\Omega \in \mathcal{F}_{\wedge}^{+}(\mathcal{H})$ and $L_0 \Omega = h_{\frac{1}{2}} \Omega$, and $m \in \mathbf{N}_0 \setminus \{1\}$, since we have an even number of particles or none and $e_{\frac{1}{2}} \wedge e_{\frac{1}{2}} = 0$. The corresponding eigenspaces are finite dimensional, $V_{h+m} = B_m$, $m \in \mathbf{N}$.

In both cases Ω is a lowest weight vector and $V_h = \text{span}\{\Omega\} = B_0 \cap \mathcal{F}_{\wedge}^{+}(\mathcal{H})$ is one dimensional.

Consider next ρ_q^{-}, where $\mathcal{K} = \mathcal{F}_{\wedge}^{-}(\mathcal{H})$ and $\mathcal{D}_\rho = \mathcal{D}_0^{-}$. Again, $L_0 = \rho_q^{-}(d_0)$ diagonalizes on $\mathcal{D}_0^{-}$, with eigenvalues of the form $h_q + m$.

If $q = 0$ then $h_0 = \frac{1}{16}$, $L_0 e_0 = h_0 e_0$ with $e_0 \in \mathcal{F}_{\wedge}^{-}(\mathcal{H})$ and $m \in \mathbf{N}_0$. Here e_0 is the lowest weight vector, $V_h = B_0 \cap \mathcal{F}_{\wedge}^{-}(\mathcal{H}) = \text{span}\{e_0\}$, and $V_{h+m} = B_m$ for $m \in \mathbf{N}$. If $q = \frac{1}{2}$, then $h_{\frac{1}{2}} = \frac{1}{2}$, since $e_{\frac{1}{2}} \in \mathcal{F}_{\wedge}^{-}(\mathcal{H})$ and $L_0 e_{\frac{1}{2}} = h_{\frac{1}{2}} e_{\frac{1}{2}}$, and $m \in \mathbf{N}_0$. Here $e_{\frac{1}{2}}$ is the lowest weight vector, $V_h = B_{\frac{1}{2}} = \text{span}\{e_{\frac{1}{2}}\}$ and $V_{h+m} = B_{m+\frac{1}{2}}$.

Hence it turns out that the representations $\rho_q^{\pm}$, $q = 0, \frac{1}{2}$, are lowest weight representations of the Virasoro algebra, with central charge $c = \frac{1}{2}$. Notice that all energies, except for the lowest are greater than one.

As promised earlier we now return to the question of showing the irreducibility of $\rho_q^{\pm}$, $q = 0, \frac{1}{2}$. It is possible, but tedious, to show directly that $L_{-i_n} \cdots L_{-i_2} \cdot L_{-i_1}$, $0 < i \leq \cdots \leq i_n$, applied to the actual lowest weight vectors, span the corresponding representation spaces. We will only sketch the proof here, since we are giving an alternative but more elegant proof below.

From $[L_0, L_k] = -k \cdot L_k$, $k \neq 0$, it follows that L_k gives zero, when applied to the lowest weight vector, for $k \in \mathbf{N}$ (otherwise its energy eigenvalue would be lower than the lowest weight, which is a contradiction). It also follows that the energy of $L_{-i_n} \cdots L_{-i_2} \cdot L_{-i_1}$ applied to the lowest weight vector is $h_q + \sum_{j=1}^{n} i_j$, where h_q denotes the energy eigenvalue of the lowest weight vector. Now, direct calculations shows that $L_{-i_n} \cdots L_{-i_2} \cdot L_{-i_1}$, $0 < i \leq \cdots \leq i_n$, applied to the lowest weight vector gives a linear combination of product basis vectors of the form $e_{k_1} \wedge \cdots \wedge e_{k_l}$, such that $\sum_{j=1}^{l} k_j = \sum_{j=1}^{n} i_j = m$. Moreover, the product basis vectors span a dense set in B_m. Hence, $L_{-i_n} \cdots L_{-i_2} \cdot L_{-i_1}$ applied to the lowest weight vector span a dense set in $\mathcal{F}_{\wedge}(\mathcal{H})$.

We now turn to the more elegant proof. Suppose that one of the representations ρ is reducible, then the corresponding representation space would have a invariant subspace not containing the actual lowest weight vector. This subspace then contains a singular vector, i.e. a vector of lowest energy of the form $h+m_n > 1$ (or, equivalently, a vector G_0 such that $L_m G_0 = 0$ for all $m \in \mathbf{N}$). This singular vector generates a unitary representation of Vir, with central charge $c = \frac{1}{2}$ and lowest weight $h + m_n > 1$, contradicting the fact that unitary representations of Vir with central charge $c = \frac{1}{2}$ have lowest weight either 0, $\frac{1}{2}$ or $\frac{1}{16}$ (see [K-R, p. 139]). This fact will be further elaborated in section 4.3.

We summarize the above discussion in the following theorem.

Theorem 29 *The representations $\rho_q^\pm : d_n \to L_n$ of the Virasoro algebra in $\mathcal{F}_\wedge^\pm(\mathcal{H})$ with common domain $\mathcal{D}_0^\pm$, as constructed above, are irreducible unitary lowest weight representations of positive and finite energy of the Virasoro algebra, with central charge $c = \frac{1}{2}$.*

In the Ramond sector, $q = 0$, with lowest weight $h_0^+ = h_0^- = \frac{1}{16}$ and corresponding lowest weight vectors Ω and e_0 respectively.

In the Neveu-Schwarz sector, $q = \frac{1}{2}$, with lowest weight $h_{\frac{1}{2}}^+ = 0$ and $h_{\frac{1}{2}}^- = \frac{1}{2}$ and corresponding lowest weight vectors Ω and $e_{\frac{1}{2}}$ respectively.

Proof. A direct consequence of the above discussion. □

At this point we have constructed some neat representations of the Virasoro algebra in the anti-symmetric Fock Hilbert space, using the fermionic oscillator algebra. Usually one constructs representations using the bosonic oscillator algebra, corresponding to the symmetric Fock Hilbert space. In the next chapter, on applications, we give some examples of representations of the Virasoro algebra, in both the symmetric and the anti-symmetric Fock Hilbert space, in the discussion of the diffeomorphisms group. That there exists representation in both cases is not surprising in view of the so-called boson-fermion correspondance, to be discussed in section 5.1, see also [K-R, p. 49-64] or [Mi, p. 193 - 202] for futher details. However, we first turn to the construction of a series of representations of the Virasoro algebra in the anti-symmetric Fock Hilbert space, in the next section, by other means.

4.4 Representations of Vir with $c_m \in [0, 1[$

There are, briefly speaking, three essentially different algebraic methods to construct unitary representations of the Virasoro algebra: the Sugawara construction, described in section 4.1 (giving representations with central charge c greater than or equal to 1); the "oscillator constructions", described in section 4.3 in the case of the fermionic oscillator algebra (giving representations with central charge $c = \frac{1}{2}$); and finally the Goddard-Kent-Olive construction, which generalizes the Sugawara construction and which we will deal with in this section (giving representations with central charge c belonging to a discrete series in $[0,1[$). Moreover, there are some "analytical" methods, described in Chap. 5, to construct unitary representations of the Virasoro algebra (with central charge $c = 1$).

For an excellent, but purely algebraic, treatment of the Goddard-Kent-Olive construction we refer to [K-R, Chap. 4 and 9-12]. As in the other cases we intend to approach the method in a self-contained manner, dealing with representations in the Fock Hilbert spaces. More precisely, we construct unitary highest weight representations of the Virasoro algebra in the tensor products of subspaces of some anti-symmetric Fock Hilbert spaces. This construction will explicitly make use of unitary highest weight representations of the Kac-Moody algebra of $\mathfrak{sl}_2$, the complex Lie algebra of traceless 2×2 matrices (which itself uses some explicitly constructed unitary highest weight representations of some particular infinite matrix Lie algebras) together with the loop and Kac-Moody algebras. This section is based on sections 4.1 and 4.2.

Let the algebraic direct sum $V = \oplus_{alg} \mathrm{span}\{v_i\}$ denotes an infinite-dimensional complex vector space with basis $\{v_i : i \in \mathbf{Z}\}$. Any vector $u \in V$ may be written as $u = (u_i)_{i \in \mathbf{Z}}$ with respect to the above basis, where only a finite number of the u_i's are non-zero, whereby we may identify V with $\mathbf{C}^\infty$. Let $\mathfrak{a}'_\infty$ denote the (Lie) algebra consisting of elements $a = (a_{i,j})_{i,j \in \mathbf{Z}}$ such that $a_{i,j} = 0$, whenever $|i - j| > N$ for some $N \in \mathbf{N}$. The elements $a = (a_{i,j})_{i,j \in \mathbf{Z}}$ are called infinite matrices and they have a finite number of non-zero diagonals, $(a_{i,i+k})_{i \in \mathbf{Z}}, k \in \mathbf{Z}$. By the ordinary matrix product, $ab = (\sum_{j \in \mathbf{Z}} a_{i,j} b_{j,k})_{i,k \in \mathbf{Z}}$, it follows that $ab \in \mathfrak{a}'_\infty$, whenever $a, b \in \mathfrak{a}'_\infty$. Notice that only a finite number of terms in the sum contribute, hence, the sum is finite. With the ordinary matrix commutator as Lie product $\mathfrak{a}'_\infty$ becomes a Lie algebra. Let $\mathfrak{gl}_\infty$ denote the subalgebra of $\mathfrak{a}'_\infty$, consisting of elements a such that only a finite number of entries are non-zero. It is a Lie algebra with respect to the ordinary matrix commutator. We notice that the unit matrices

$e_{i,j}$, $i,j \in \mathbf{Z}$, with 1 entry at the (i,j)'th position and zero entries elsewhere, form a basis for $\mathfrak{gl}_\infty$. Moreover

$$e_{i,j}e_{m,n} = \delta_{j-m}e_{i,n}$$

and

$$[e_{i,j}, e_{m,n}] = \delta_{j-m}e_{i,n} - \delta_{i-n}e_{m,j}$$

There is a natural action (representation) of $\mathfrak{a}'_\infty$ and of $\mathfrak{gl}_\infty$ on V in terms of the unit matrices, given by

$$e_{i,j}v_k = \delta_{j-k}v_i$$

Then the shift operator s_k on V, $k \in \mathbf{Z}$, given by

$$s_k v_i = v_{i-k}$$

may be written as

$$s_k = \sum_{i \in \mathbf{Z}} e_{i,i+k}$$

Therefore, s_k acts as the matrix with 1 entry on the k'th diagonal and zeros elsewhere, hence s_k corresponds to an element in $\mathfrak{a}'_\infty$ which we also denote s_k. The elements s_k, $k \in \mathbf{Z}$, form a commutative subalgebra of $\mathfrak{a}'_\infty$, since $[s_k, s_l] = 0$, for $k, l \in \mathbf{Z}$. We also notice that $\{s_k : k \in \mathbf{Z}\}v_i = \{v_j : j \in \mathbf{Z}\}$, for any $i \in \mathbf{Z}$.

Let $\mathfrak{gl}_2$ denote the Lie algebra of all 2×2 matrices with complex entries and $\mathfrak{sl}_2$ the Lie subalgebra of all traceless complex 2×2 matrices. At this point we notice that one could easily consider the corresponding $n \times n$ matrices instead, for any $n \in \mathbf{N}$, and that the following considerations also hold in general. However, we consider only the case $n = 2$, since it is sufficient for our purpose. Of course, both the Lie algebras above have natural actions on $\mathbf{C}^2$.

Let $\widetilde{\mathfrak{gl}}_2$ and $\widetilde{\mathfrak{sl}}_2$ be the loop algebras, as defined in section 4.1, corresponding to $\mathfrak{gl}_2$ and $\mathfrak{sl}_2$, repectively. If the matrix units for $\mathfrak{gl}_2$ and $\mathfrak{sl}_2$ are denoted by $e_{i,j}$, $i,j \in \mathbf{Z}$ (they are 2×2 matrices, but they may be identified canonical with $e_{i,j}$, $i,j \in \mathbf{Z}$ in $\mathfrak{gl}_\infty$, why we use the same symbol) then

$$e_{i,j}(k) = e_{i,j} \otimes t^k$$

for $i,j = 1,2$ and $k \in \mathbf{Z}$, form a basis for $\widetilde{\mathfrak{gl}}_2$, as $e_{i,j}$, $i,j = 1,2$, for $\mathfrak{gl}_2$. The multiplication in $\widetilde{\mathfrak{gl}}_2$ is given by

$$e_{i,j}(k)e_{m,n}(l) = e_{i,j}e_{m,n} \otimes t^{k+l} = \delta_{j-m}e_{i,n}(k+l)$$

on basis elements, whereby $\tilde{\mathfrak{gl}}_2$ forms an associative algebra. The Lie bracket on $\tilde{\mathfrak{gl}}_2$ becomes

$$[e_{i,j}(k), e_{m,n}(l)] = \delta_{j-m} e_{i,n}(k+l) - \delta_{i-n} e_{n,j}(k+l)$$

as intended, see section 4.1.

Consider now the natural action of $\mathfrak{gl}_2$ on $\mathbf{C}^2$ with standard basis $\{u_1, u_2\}$. Then the loop algebra $\tilde{\mathfrak{gl}}_2$ acts naturally in $\mathbf{C}[t, t^{-1}]^2 \cong \mathbf{C}^2 \otimes \mathbf{C}[t, t^{-1}]$, where $\mathbf{C}[t, t^{-1}]$, as previously, denotes the ring of Laurent polynomials, as follows. The vectors

$$v_{2k+j} = u_j \otimes t^{-k}$$

form a basis of $\mathbf{C}[t, t^{-1}]^2$ over $\mathbf{C}$, where the index $i = 2k + j \in \mathbf{Z}$, since $k \in \mathbf{Z}$ and $j = 1, 2$. Hereby we identify $\mathbf{C}[t, t^{-1}]^2$ with $\mathbf{C}^\infty$. Since $e_{i,j}(k) = e_{i,j} \otimes t^k$ and $v_{2n+j} = u_j \otimes t^{-n}$ it follows that

$$e_{i,j}(k)v_{2n+j} = (e_{i,j}u_j) \otimes t^{k-n} = u_i \otimes t^{k-n} = v_{2(k-n)+i}$$

This action allow us to determine the corresponding matrix $\tau(a(t))$ in $\mathfrak{a}'_\infty$ of any element $a(t) \in \tilde{\mathfrak{gl}}_2$. For the basis elements it is simply defined as

$$\tau(e_{i,j}(k)) = \sum_{n \in \mathbf{Z}} e_{2(n-k)+i, 2n+j}$$

where $e_{2(n-k)+i, 2n+j}$ denotes the matrix units in $\mathfrak{a}'_\infty$ and $e_{i,j}$ those in $\mathfrak{gl}_2$ ($n, k \in \mathbf{Z}$ and $i, j = 1, 2$). For a general $a(t) \in \tilde{\mathfrak{gl}}_2$ it follows that the corresponding matrix in $\mathfrak{a}'_\infty$ takes the block form

$$\tau(a(t)) = \begin{pmatrix} \ddots & \ddots & \ddots & & \\ & a_{-1} & a_0 & a_1 & \\ & & a_{-1} & a_0 & a_1 \\ & & & a_{-1} & a_0 & a_1 \\ & & & & \ddots & \ddots & \ddots \end{pmatrix}$$

where $a(t) = \sum_{k \in \mathbf{Z}} a_k \otimes t^k$ and each $a_k \in \mathfrak{gl}_2$. Thus τ is an injective homomorphism of $\tilde{\mathfrak{gl}}_2$ as an associative algebra and, thus, also as a Lie algebra. The image of $a(t) = \sum_{k \in \mathbf{Z}} a_k \otimes t^k$ under τ is evidently a strictly upper triangular matrix if and only if $a_k = 0$, for $k \in -\mathbf{N}$ and a_0 strictly upper triangular. Moreover, if a^* denotes the usual

Hermitian adjoint of the 2×2 matrix a and we define an anti-linear anti-involution ω on $\widetilde{\mathfrak{gl}}_2$ by

$$\omega(a(k)) = a^* \otimes t^{-k} = a^*(-k)$$

then it follows directly that

$$\tau(\omega(a(k))) = \tau(a(k))^*$$

where $\tau(a(k))^*$ denotes the usual Hermitian adjoint of $\tau(a(k))$ in $\mathfrak{a}'_\infty$. Furthermore

$$
\begin{aligned}
s_k = \tau\left((e_{1,2} + e_{2,1} \otimes t)^k\right) &= (\tau(e_{1,2}) + \tau(e_{2,1} \otimes t))^k \\
&= \left(\sum_{n \in \mathbf{Z}} e_{2n+1,2n+2} + \sum_{n \in \mathbf{Z}} e_{2(n-1)+2,2n+1}\right)^k \\
&= \left(\sum_{n \in \mathbf{Z}}(e_{2n+1,2n+2} + e_{2n,2n+1})\right)^k \\
&= \left(\sum_{n \in \mathbf{Z}} e_{n,n+1}\right)^k = s_1^k = s_k
\end{aligned}
$$

Suppose there exists a projective representation of $\mathfrak{a}'_\infty$ with two-cocycle $\alpha(a,b), a,b \in \mathfrak{a}'_\infty$. Then this can be made into a linear representation of the central extension $\mathfrak{a}_\infty$ of $\mathfrak{a}'_\infty$, i.e.

$$\mathfrak{a}_\infty = \mathfrak{a}'_\infty \oplus \mathbf{C} \cdot c$$

with c in the center of $\mathfrak{a}_\infty$ and the new bracket given by

$$[a,b] = [a,b]_0 + \alpha(a,b) \cdot c$$

where $[a,b]_0 = ab - ba$ now denotes the usual commutator in $\mathfrak{a}'_\infty$. Since $\widetilde{\mathfrak{gl}}_2$ can be realized as a subalgebra of $\mathfrak{a}'_\infty$ there is also a projective representation of $\widetilde{\mathfrak{gl}}_2$, by the restriction (we suppose that there is one of $\mathfrak{a}'_\infty$). Then there is also a linear representation of the central extension $\widehat{\mathfrak{gl}}_2 = \widetilde{\mathfrak{gl}}_2 \oplus \mathbf{C} \cdot c$, as a subalgebra of $\mathfrak{a}_\infty$. Moreover, since $\mathfrak{sl}_2$ is a Lie subalgebra of $\mathfrak{gl}_2$ we may define $\widetilde{\mathfrak{sl}}_2$ and $\widehat{\mathfrak{sl}}'_2$ as subalgebras of $\widetilde{\mathfrak{gl}}_2$ and $\widehat{\mathfrak{gl}}'_2$, respectively, in an obvious manner. Then we may consider $\widehat{\mathfrak{sl}}'_2$ as a subalgebra of $\mathfrak{a}_\infty$ too. We emphasize that $s_k = \tau\left((e_{1,2}(0) + e_{2,1}(1))^k\right) \in \tau(\widetilde{\mathfrak{gl}_2})$. But $s_{2k+1} \in \tau(\widetilde{\mathfrak{sl}_2})$ and $s_{2k} \notin \tau(\widetilde{\mathfrak{sl}_2})$, since

$$(e_{1,2}(0) + e_{2,1}(1))^{2k} \cong \begin{pmatrix} 0 & 1 \\ t & 0 \end{pmatrix}^{2k} = t^k \cdot I \cong I \otimes t^k \notin \widetilde{\mathfrak{sl}_2}$$

and

$$(e_{1,2}(0) + e_{2,1}(1))^{2k+1} \cong t^k \cdot \begin{pmatrix} 0 & 1 \\ t & 0 \end{pmatrix} \in \widetilde{\mathfrak{sl}_2}$$

This will be important later on. However, we first have to discuss the representations of $\mathfrak{a}'_\infty$ supposed above.

Let $\mathcal{F}_\wedge(\mathcal{H})$ denote the anti-symmetric Fock Hilbert space, where $\{e_i : i \in \mathbf{Z}\}$ denotes an orthonormal basis for the one particle Hilbert space $\mathcal{H}$ (e_i with $i \in -\mathbf{N}$ describing the fermionic anti-particles).

Define operators $E_{i,j}, i, j \in \mathbf{Z}$, on $\mathcal{F}_\wedge(\mathcal{H})$ by

$$E_{i,j} = \begin{cases} a^*(e_i)a(e_j) & \text{, for } i, j \geq 0 \\ a^*(e_i)a^*(e_j) & \text{, for } i \geq 0 > j \\ a(e_i)a(e_j) & \text{, for } j \geq 0 > i \\ a(e_i)a^*(e_j) - \delta_{i-j} = -a^*(e_j)a(e_i) & \text{, for } 0 > i, j \end{cases}$$

where $a^*(e_k)$ and $a(e_k)$ denote the creation and annihilation operators, respectively, in the Fock representation, described in section 2.2. For all $i, j \in \mathbf{Z}$, $E_{i,j}$ is obviously well-defined, in fact, it is a partial isometry on $\mathcal{F}_\wedge(\mathcal{H})$.

The commutator $[E_{i,j}, E_{m,n}]$ can be calculated directly using the CAR. For $i, j, m, n \geq 0$ we get

$$\begin{aligned} E_{i,j}E_{m,n} &= a^*(e_i)a(e_j)a^*(e_m)a(e_n) \\ &= a^*(e_i)\left(\delta_{j-m} - a^*(e_m)a(e_j)\right)a(e_n) \\ &= \delta_{j-m}a^*(e_i)a(e_n) - a^*(e_m)a^*(e_i)a(e_n)a(e_j) \\ &= E_{i,n}\delta_{j-m} - a^*(e_m)\left(\delta_{i-n} - a(e_n)a^*(e_i)\right)a(e_j) \\ &= E_{i,n}\delta_{j-m} - E_{m,j}\delta_{i-n} + E_{m,n}E_{i,j} \end{aligned}$$

so $[E_{i,j}, E_{m,n}] = E_{i,n}\delta_{j-m} - E_{m,j}\delta_{i-n}$. In a completely identical manner one obtains exactly the same commutator relations for all other cases except for $i \geq 0 > j$ and $n \geq 0 > m$, and for $j \geq 0 > i$ and $m \geq 0 > n$. For these two exceptional cases one obtains

$$[E_{i,j}, E_{m,n}] = E_{i,n}\delta_{j-m} - E_{m,j}\delta_{i-n} \pm \delta_{j-m}\delta_{i-n}$$

where the $+$ sign holds for j (and m) non-negative and the $-$ sign holds for j (and m) negative. The results are obtained by calculations similar to those above, for example

$$
\begin{aligned}
E_{i,j} E_{m,n} &= a^*(e_i) a^*(e_j) a(e_m) a(e_n) \\
&= a^*(e_i) \left(\delta_{j-m} - a(e_m) a^*(e_j) \right) a(e_n) \\
&= \delta_{j-m} a^*(e_i) a(e_n) - a(e_m) a^*(e_i) a(e_n) a^*(e_j) \\
&= E_{i,n} \delta_{j-m} - a(e_m) \left(\delta_{i-n} - a(e_n) a^*(e_i) \right) a^*(e_j) \\
&= E_{i,n} \delta_{j-m} - E_{m,j} \delta_{i-n} - \delta_{j-m} \delta_{i-n} + E_{m,n} E_{i,j}
\end{aligned}
$$

for $i \geq 0 > j$ and $n \geq 0 > m$, from where the desired commutation relations follows. Define α on $\mathfrak{a}'_\infty$ by bilinearity and

$$
\begin{aligned}
\alpha(e_{i,j}, e_{m,n}) &= \begin{cases} 1 & , \text{ for } j = m \geq 0 \text{ and } i = n < 0 \\ -1 & , \text{ for } j = m < 0 \text{ and } i = n \geq 0 \\ 0 & , \text{ otherwise} \end{cases} \\
&= \delta_{j-m} \delta_{i-n} \left(\chi(-i) \chi(j+1) - \chi(-j) \chi(i+1) \right)
\end{aligned}
$$

on the matrix units, where χ denotes the indicator function for $\mathbf{N}$. Thereby $\alpha(\cdot, \cdot)$ defines a two-cocycle sometimes called the Kac-Peterson two-cocycle ([Mi, p 179]). Then we get a projective representation π of $\mathfrak{a}'_\infty$ in $\mathcal{F}_\wedge(\mathcal{H})$ by putting $\pi(e_{i,j}) = E_{i,j}$ together with linearity. Thus

$$
[\pi(e_{i,j}) \pi(e_{m,n})] = \pi \left([e_{i,j}, e_{m,n}] \right) + \alpha(e_{i,j}, e_{m,n}) \cdot I
$$

Of couse we have to check that the extension of π to all of $\mathfrak{a}'_\infty$ is well-defined by linearity. Notice first that any element in $\mathfrak{a}'_\infty$ may be written as a finite linear combination of elements of the form

$$
a_k = \sum_{i \in \mathbf{Z}} \lambda_i e_{i,i+k}
$$

due to the fact that any element of $\mathfrak{a}'_\infty$ has a finite number of non-vanishing diagonal entries. The above $\lambda_i \in \mathbf{C}$ are arbitrary. So the extension becomes well-defined on $\mathfrak{a}'_\infty$ if there is a well-defined extension to elements of the form a_k. Observe that $\sum_{i \in \mathbf{Z}} \lambda_i E_{i,i+k}$ can be split into four sum-terms corresponding to $i \in \{j \in \mathbf{N} \cup \{0\} : j \geq -k\}$, $i \in \{j \in \mathbf{N} \cup \{0\} : j < -k\}$ which contribute only for $k < 0$, $i \in \{j \in -\mathbf{N} : j < -k\}$ and $i \in \{j \in -\mathbf{N} : j \geq -k\}$ which contribute only for $k > 0$. Each of these sum-terms reduces to a finite sum on Ω and on basis product vectors (where we utilize the term $-\delta_{i-j}$, for $i = j < 0$). In particular for $i = i + k < 0$ (i.e. $k = 0$) we get

$$\sum_{i \in \mathbf{Z}} \lambda_i E_{i,i}(e_{j_1} \wedge \cdots \wedge e_{j_n}) = \sum_{i \in \{j_1, \ldots, j_n\}} (-\lambda_i)(e_{j_1} \wedge \cdots \wedge e_{j_n})$$

on an arbitrary basis product vector (in this section we use the convention that $j_1 > \cdots > j_n$ for product vectors). Hence, we may extend the representation π to all of $\mathfrak{a}'_\infty$ in a well-defined manner by linearity, since the formally infinite sum reduces to a finite sum.

As mentioned earlier we may turn the projective representation constructed above into a linear Lie algebra representation. Define the central extension $\mathfrak{a}_\infty = \mathfrak{a}'_\infty \oplus \mathbf{C} \cdot c$ of $\mathfrak{a}'_\infty$, where c denotes a central element, with the new bracket

$$[a, b] = [a, b]_0 + \alpha(a, b) \cdot c$$

where $[a, b]_0 = ab - ba$ now denotes the old bracket on $\mathfrak{a}'_\infty$ given by the ordinary commutator. We extend π from $\mathfrak{a}'_\infty$ to $\mathfrak{a}_\infty$ by putting $\pi(c) = 1$. Thus, we have a Lie algebra representation of $\mathfrak{a}_\infty$ in $\mathcal{F}_\wedge(\mathcal{H})$ given by π, constructed above. The Hilbert space inner product $\langle \cdot, \cdot \rangle$ then defines a contravariant Hermitian form, since $\langle \pi(a)F, G \rangle = \langle F, \pi(a^*)G \rangle$, for all $F, G \in \mathcal{F}_\wedge(\mathcal{H})$, where a^* denotes the earlier defined anti-linear anti-involution on $\mathfrak{a}'_\infty$. Thus, π is a unitary representation since the inner product is positively definite, by definition.

The charge operator Q is defined by $Q\Omega = 0$ and

$$Q(e_{j_1} \wedge \cdots \wedge e_{j_n}) = \left(\sum_{i=1}^n j_i\right)(e_{j_1} \wedge \cdots \wedge e_{j_n})$$

on basis vectors. It extends to all of $\mathcal{D}$ by linearity, where $\mathcal{D}$ denotes $\oplus_{alg} \wedge^n \mathcal{H}$ (it is even well-defined on $\mathcal{D}(N)$, where N denotes the number operator). Considerations such as those in section 2.3 show that Q has spectrum $\mathbf{Z}$ and that $\mathcal{F}_\wedge(\mathcal{H})$ may be decomposed as

$$\mathcal{F}_\wedge(\mathcal{H}) = \oplus_{q \in \mathbf{Z}} \mathcal{H}_q$$

where $\mathcal{H}_q$ denotes the eigenspace of Q corresponding to eigenvalue $q \in \mathbf{Z}$. Obviously, each $E_{i,j}$ conserves the charge of any basis product vector (or yields zero), hence, it leaves each charge sector $\mathcal{H}_q$

invariant. Thereby $\pi_q = \pi|_{\mathcal{H}_q}$ defines a unitary representation of $\mathfrak{a}'_\infty$ in $\mathcal{H}_q$, for each $q \in \mathbf{Z}$.

Define $\mathbf{B}_0 = \mathrm{span}\{\Omega, e_0\}$ and

$$\mathbf{B}_m = \mathrm{span}\{e_{j_1} \wedge \cdots \wedge e_{j_n} : \sum_{i=1}^{n} |j_i| = m, n \in \mathbf{N}\}$$

for any $m \in \mathbf{N}$. Then we may decompose $\mathcal{F}_\wedge(\mathcal{H})$ as a Hilbert space direct sum

$$\mathcal{F}_\wedge(\mathcal{H}) = \oplus_{m \in \mathbf{N} \cup \{0\}} \mathbf{B}_m$$

Moreover, we define $\mathcal{D}_0$ as the algebraic direct sum

$$\mathcal{D}_0 = \oplus_{alg} \mathbf{B}_m$$

Thus $\mathcal{D}_0$ is a dense subspace of $\mathcal{F}_\wedge(\mathcal{H})$. We now define an energy operator H on $\mathcal{D}_0$ by

$$H F_m = m F_m$$

for any $F_m \in \mathbf{B}_m$ and by linearity. Notice that the spectrum of H indeed is $\mathbf{N} \cup \{0\}$ and that $\mathcal{F}_\wedge(\mathcal{H})$ decomposes into a direct Hilbert space sum of eigenspaces of H corresponding to eigenvalues $m \in \mathbf{N} \cup \{0\}$. Observe that $H\Omega = 0$ and

$$H(e_{j_1} \wedge \cdots \wedge e_{j_n}) = \left(\sum_{i=1}^{n} |j_i|\right)(e_{j_1} \wedge \cdots \wedge e_{j_n})$$

on basis vectors. We define the sector vacuum Ω_q in each sector $\mathcal{H}_q$ to be a distinguished vector, namely the unique vector with lowest energy $m = \frac{1}{2}q(q-1)$, i.e.

$$\Omega_q = \begin{cases} e_{q-1} \wedge \cdots \wedge e_1 \wedge e_0 & , \text{ for } q > 0 \\ e_{-1} \wedge e_{-2} \wedge \cdots \wedge e_q & , \text{ for } q < 0 \\ \Omega & , \text{ for } q = 0 \end{cases}$$

A direct calculation shows that

$$H\Omega_q = \frac{1}{2}q(q-1) \cdot \Omega_q$$

(the calculations are similar to those we will present in detail later on, see section 5.2 for a particular case). Moreover, it follows by straight-forward calculations that

$$E_{i_1,j_1} \cdot \ldots \cdot E_{i_n,j_n} \Omega_q$$

span a dense set in $\mathcal{H}_q$.

Below we will need the following decomposition of each $\mathcal{H}_q$ into sector-energy subspaces

$$\mathcal{H}_q = \oplus_{m \in \mathbf{N} \cup \{0\}} \mathcal{H}_q^{(m)}$$

where $\mathcal{H}_q^{(m)}$ denotes the eigenspace of the sector Hamiltonian $H|_{\mathcal{H}_q} - \frac{1}{2}q(q-1)$ corresponding to the sector-energy eigenvalue m, which equals the eigenspace of the energy operator corresponding to energy eigenvalue $m + \frac{1}{2}q(q-1)$, i.e.

$$\mathcal{H}_q^{(m)} = \mathcal{H}_q \cap \mathbf{B}_{m+\frac{1}{2}q(q-1)}$$

for each $q \in \mathbf{Z}$.

We know that each $\mathcal{H}_q$ is invariant under $\pi(\mathfrak{a}'_\infty)$ and that $\pi(\mathfrak{a}'_\infty)\Omega_q$ form a dense set in $\mathcal{H}_q$. Then it follows that π is irreducible. Suppose that U is an invariant subspace of $\mathcal{H}_q$, then $\mathcal{H}_q = U \oplus U^\perp$, where the orthogonal complement $U^\perp$ of U is taken in $\mathcal{H}_q$. Then the decomposition $\mathcal{H}_q = U \oplus U^\perp$ also respects the sector-energy decomposition which can be proved quite similarly to the proof of theorem 22 part 3, i.e. we may write

$$U = \oplus_{m \in \mathbf{N} \cup \{0\}} U_m \qquad \text{and} \qquad U^\perp = \oplus_{m \in \mathbf{N} \cup \{0\}} U_m^\perp$$

where $U_m = U \cap \mathcal{H}_q^{(m)}$ and $U_m^\perp = U^\perp \cap \mathcal{H}_q^{(m)}$. Since $\mathcal{H}_q^{(0)} = \text{span}\{\Omega_q\}$ is one-dimensional, Ω_q belongs to either U_0 or $U_0^\perp$, say $\Omega_q \in U_0$, but Ω_q generates a dense set in $\mathcal{H}_q$ under $\pi(\mathfrak{a}'_\infty)$. Due to the invariance of $U = \oplus_{m \in \mathbf{N} \cup \{0\}} U_m$ it follows that $U = \mathcal{H}_q$ and $U^\perp = \{0\}$. Hence, $\pi_q = \pi|_{\mathcal{H}_q}$ is irreducible.

Observe that

$$E_{i,j} \mathcal{H}_q^{(m)} \subset \mathcal{H}_q^{(m+i-j)}$$

which easily follows by checking the four possible cases of $E_{i,j}$ directly. This inspires us to define the degree of $e_{i,j} \in \mathfrak{gl}_\infty$ as

$$\deg(e_{i,j}) = i - j$$

Futhemore, we define the degree of an element $a \in \mathfrak{gl}_\infty$ to be $\deg(a) = k$ if a is of the form $a = \sum \lambda_i e_{i,i+k}$, where the sum is over a finite subset of $\mathbf{Z}$. We may decompose $\mathfrak{gl}_\infty$ as a direct sum of homogeneous components $\mathfrak{g}_k$, i.e. subspaces of elements in $\mathfrak{gl}_\infty$ with fixed degree $k \in \mathbf{Z}$, as $\mathfrak{gl}_\infty = \oplus_{k \in \mathbf{Z}} \mathfrak{g}_k$. It then follows that

$$\pi(\mathfrak{g}_k) \mathcal{H}_q^{(m)} \subset \mathcal{H}_q^{(m+k)}$$

and
$$\pi(\mathfrak{g}_k)\Omega_q = 0$$

whenever $k < 0$, since $E_{i,j}\Omega_q = 0$, for $i < j$. Since $E_{i,j} = \pi(e_{i,j}) \in \pi(\mathfrak{g}_{j-i})$ it follows that

$$\sum_{\substack{k_1+\cdots+k_n=m \\ k_1,\ldots,k_n \in \mathbb{N}\cup\{0\}}} \pi_q(\mathfrak{g}_{k_1})\cdot\ldots\cdot\pi_q(\mathfrak{g}_{k_n})\Omega_q$$

form a dense set in $\mathcal{H}_q^{(m)}$.

Let
$$\mathfrak{n}_+ = \oplus_{k<0}\mathfrak{g}_k$$

such that any $a \in \mathfrak{n}_+$ is a strictly upper triangular matrix in $\mathfrak{gl}_\infty$. Then
$$\pi_q(\mathfrak{n}_+)\Omega_q = 0$$

and
$$\pi_q(e_{i,i})\Omega_q = E_{i,i}\Omega_q = \lambda_i\Omega_q$$

with
$$\lambda_i = \begin{cases} 1 & \text{, for } 0 \le i \le q-1 \quad (q>0) \\ -1 & \text{, for } q \le i < 0 \qquad (q<0) \\ 0 & \text{, otherwise} \end{cases} \tag{4.2}$$

Summarizing the above we get

Theorem 30 *For all $q \in \mathbf{Z}$, π_q defines an irreducible unitary highest weight representation of $\mathfrak{gl}_\infty$ in $\mathcal{H}_q$ with the sector vacuum vector Ω_q as the highest weight vector and $\lambda(q) = \{\lambda_i : i \in \mathbf{Z}\}$, where λ_i is defined by formula (4.2), as the highest weight.*

Proof. Notice that our definition of a highest weight representation follows that in [K-R, p 41]; a collection of numbers $\lambda = \{\lambda_i : i \in \mathbf{Z}\}$ is called a highest weight of an irreducible representation π_λ in a vector space $\mathcal{L}(\lambda)$, called a highest weight representation, if there exists a non-zero vector ν_λ in $\mathcal{L}(\lambda)$, called a highest weight vector, such that
$$\pi_\lambda(\mathfrak{n}_+)\nu_\lambda = 0$$

and
$$\pi_\lambda(e_{i,i})\nu_\lambda = \lambda_i\nu_\lambda$$

We note that $\mathcal{L}(\lambda)$ is determined by λ. Let $\lambda = \lambda(q)$, $\mathcal{L}(\lambda) = \mathcal{H}_q$, $\nu_\lambda = \Omega_q$, which is allowed due to the unique correspondance between

q and $\{\lambda_i : i \in \mathbf{Z}\}$, given in (4.2). Thus, it follows directly from the above that π_q defines an irreducible highest weight representation of $\mathfrak{gl}_\infty$ in $\mathcal{H}_q$ with highest weight vector Ω_q and highest weight $\lambda(q)$. The unitarity of π_q follows, since the representation π_q of $\mathfrak{a}'_\infty$ is unitary. □

The tensor product of two irreducible unitary highest weight representations with highest weight vectors ν_1 and ν_2, respectively, gives a irreducible unitary highest weight representation in the *highest component* of the tensor product of the representation spaces, i.e. in the vector space generated by the highest weight vector $\nu_1 \otimes \nu_2$. Therefore we may construct irreducible unitary highest weight representations π_λ of $\mathfrak{gl}_\infty$ in a vector space $\mathcal{L}(\lambda)$ with highest weight $\lambda = k_1 \cdot \lambda(q_1) + \cdots + k_n \cdot \lambda(q_n)$, and highest weight vector $\nu_\lambda = \Omega_{q_1}^{\otimes k_1} \otimes \cdots \otimes \Omega_{q_n}^{\otimes k_n}$, where $\Omega_{q_j}^{\otimes k_j} = \Omega_{q_j} \otimes \cdots \otimes \Omega_{q_j}$, k_j times, $j = 1, \ldots, n$, $k_1, \ldots, k_n \in \mathbf{N}$, $q_1, \ldots, q_n \in \mathbf{Z}$ and $n \in \mathbf{N}$.

Returning for a while to the commutative subalgebra spanned by the shift operators in $\mathfrak{a}'_\infty$ we have

$$[\pi_q(s_n), \pi_q(s_k)] = \alpha(s_n, s_k) \cdot I$$

where

$$
\begin{aligned}
\alpha(s_n, s_k) &= \sum_{i \in \mathbf{Z}} \sum_{j \in \mathbf{Z}} \alpha(e_{i,i+n}, e_{j-k,j}) \\
&= \sum_{i \in \mathbf{Z}} \sum_{j \in \mathbf{Z}} \delta_{i-j} \delta_{i+n-(j-k)} \left(\chi(-i)\chi(i+n+1) - \chi(i+1)\chi(-i-n) \right) \\
&= \sum_{-n \le i < 0} \delta_{n+k} - \sum_{0 \le i < -n} \delta_{n+k} \\
&= n \cdot \delta_{n+k}
\end{aligned}
$$

One recognizes this as the commutation relations of the fermionic oscillator algebra, also known as the Heisenberg algebra (see for example [K-R, p 12]). Furthermore

$$\pi_q(s_0) = \sum_{i \in \mathbf{Z}} \pi_q(e_{i,i}) = \sum_{i \in \mathbf{Z}} E_{i,i} = q \cdot I$$

when applied to any vector in $\mathcal{H}_q$ and

$$\pi_q(s_k)\Omega_q = 0$$

for $k > 0$. Moreover, the vectors

$$\pi_q(s_{-k_1}) \cdot \ldots \cdot \pi_q(s_{-k_n})\Omega_q$$

where $k_1 \geq \cdots \geq k_n > 0$, are linearly independent and those with $\sum_{j=1}^n k_j = m$ belong to $\mathcal{H}_q^{(m)}$, in fact, they form a basis for $\mathcal{H}_q^{(m)}$. This can be seen as follows. There are obviously $p(m)$ linearly independent vectors of the given form in $\mathcal{H}_q^{(m)}$, where $p(m)$ denotes the number of partitions of $m \in \mathbf{N}$ into a sum of positive integers $\sum_{i=1}^n k_i = m$. But $p(m)$ is the dimension of $\mathcal{H}_q^{(m)}$ (this is shown in detail in the proof of theorem 34 in section 5.1, for a similar case). So there are $\dim(\mathcal{H}_q^{(m)})$ numbers of linearly independent vectors and thus they form a basis of $\mathcal{H}_q^{(m)}$. Then π_q is an irreducible representation of the commutative subalgebra, since each $\mathcal{H}_q$ is indeed invariant under $\{\pi_q(s_k) : k \in \mathbf{N}\}$ and the linearly independent vectors $\pi_q(s_{-k_1}) \cdot \ldots \cdot \pi_q(s_{-k_n})\Omega_q$, with $k_1 \geq \cdots \geq k_n > 0$ span a dense set of $\mathcal{H}_q$, hence, the irreducibility follows by a proof similar to that used in the case of $\mathfrak{a}'_\infty$ above.

We now return to the the loop algebras. Following the approach of section 4.1 we may, as mentioned earlier, extend the loop algebra $\widetilde{\mathfrak{gl}}_2$ with a central element c yielding

$$\widehat{\mathfrak{gl}}'_2 = \widetilde{\mathfrak{gl}}_2 \oplus \mathbf{C} \cdot c$$

with respect to the Kac-Peterson two-cocycle, which becomes

$$\alpha\left(\tau(e_{i,j}(k)), \tau(e_{m,n}(l))\right) = \delta_{i-n}\delta_{j-m}\delta_{k+l} \cdot k$$

By linearity we obtain

$$\alpha\left(\tau(a(k)), \tau(b(l))\right) = \delta_{k+l} \cdot k \cdot \mathrm{Tr}(ab)$$

for arbitrary $a, b \in \mathfrak{gl}_2$ and for $a(t), b(t) \in \widetilde{\mathfrak{gl}}_2$ we get

$$\alpha\left(\tau(a(t), \tau(b(t)))\right) = \mathrm{res}_0\left(\mathrm{Tr}(a'(t)b(t))\right)$$

where $a'(t)$ denotes the derivative of $a(t)$ and res_0 the residue at $t = 0$.

The Lie bracket on $\widehat{\mathfrak{gl}}'_2$ is then given by

$$[a(k), c] = 0$$

and

$$[a(k), b(l)] = [a, b]_0(k+l) + k \cdot \delta_{k+l} \cdot \mathrm{Tr}(ab) \cdot c$$

for arbitrary $a(k), b(l) \in \widetilde{\mathfrak{gl}}_2$ where $[a, b]_0 = ab - ba$, as earlier.

The Killing form $(a, b)_0 = \mathrm{Tr}(ab)$ on $\mathfrak{gl}_2$ gives a bilinear form on $\widetilde{\mathfrak{gl}}_2$ by

$$(a(k), b(l)) = \delta_{k+l}\mathrm{Tr}(ab)$$

which extend to all of $\widetilde{\mathfrak{gl}}_2$ by linearity as

$$(a(t), b(t)) = \mathrm{res}_0\left(t^{-1}\mathrm{Tr}(a(t)b(t))\right)$$

It is evidently a symmetric, invariant, non-degenerated bilinear form on $\widetilde{\mathfrak{gl}}_2$, which is verified by direct calculations. The form extends to $\widehat{\mathfrak{gl}}_2'$ by putting $(c, \widetilde{\mathfrak{gl}}_2) = 0$ and $(c, c) = 0$. On $\widehat{\mathfrak{gl}}_2'$ the form is of course degenerated, but it is still symmetric, invariant and bilinear. As in section 4.1 we therefore extend $\widehat{\mathfrak{gl}}_2'$ further, by a generator d, by

$$\widehat{\mathfrak{gl}}_2 = \widehat{\mathfrak{gl}}_2' \oplus \mathbf{C} \cdot d$$

with the old commutator on $\widehat{\mathfrak{gl}}_2'$ together with

$$[d, c] = 0$$

and

$$[d, a(k)] = k \cdot a(k)$$

giving an affine Kac-Moody algebra. It follows that $\widehat{\mathfrak{gl}}_2$ carries a non-degenerated, symmetric, invariant bilinear form $(\cdot, \cdot)$ given as in proposition 20 of section 4.1, with $s = 0$ (and $(a(t), b(t))_0 = \mathrm{Tr}(a(t)b(t))$).

Recall that $\widetilde{\mathfrak{gl}}_2$ is a subalgebra of $\mathfrak{a}_\infty'$ and that the anti-linear anti-involution ω on $\widetilde{\mathfrak{gl}}_2$ coincide with the one induced from $\mathfrak{a}_\infty'$. If we define $\omega(c) = c$, then the central extension $\widehat{\mathfrak{gl}}_2'$ of $\widetilde{\mathfrak{gl}}_2$ is a subalgebra of the central extension $\mathfrak{a}_\infty$ of $\mathfrak{a}_\infty'$, too. We define $\tau(c) = c$. Since $s_k = \tau((e_{1,2}(0) + e_{2,1}(1))^k) \in \tau(\widehat{\mathfrak{gl}}_2')$ and π_q remains irreducible when restricted to the subalgebra spanned by the s_k, $k \in \mathbf{Z}$, it follows that $\pi_q(\tau(\cdot))$ defines an irreducible representation of $\widehat{\mathfrak{gl}}_2'$ in $\mathcal{H}_q$. Moreover

$$\pi_q(\tau(a(t)))\Omega_q = 0$$

whenever $a(t) = \sum_{k \geq 0} a_k \otimes t^k$ with a_0 a strictly upper triangular matrix, and

$$\pi_q(\tau(e_{i,i}(0)))\Omega_q = \sum_{k \in \mathbf{Z}} E_{2k+i,2k+i}\Omega_q = \left(\sum_{k \in \mathbf{Z}} \lambda_{2k+i}\right) \cdot \Omega_q$$

where

$$\lambda_j = \lambda_{2k+i} = \begin{cases} 1 & \text{, for } 0 \leq j = 2k+i < q \\ -1 & \text{, for } q \leq j = 2k+i < 0 \\ 0 & \text{, otherwise} \end{cases}$$

As earlier $\pi_q(\tau(c)) = 1$.

Now we are ready to discuss the representations of $\widehat{\mathfrak{sl}}_2$ in detail. Recall that the loop algebra $\widetilde{\mathfrak{sl}}_2$, the Kac-Moody algebras $\widehat{\mathfrak{sl}}_2' = \widetilde{\mathfrak{sl}}_2 \oplus \mathbf{C} \cdot c$ and $\widehat{\mathfrak{sl}}_2 = \widehat{\mathfrak{sl}}_2' \oplus \mathbf{C} \cdot d$ are Lie subalgebras of, and defined similar to, the corresponding Lie algebras $\widetilde{\mathfrak{gl}}_2$, $\widehat{\mathfrak{gl}}_2'$ and $\widehat{\mathfrak{gl}}_2$, respectively. Let

$$x = e_{1,2} \quad , \quad y = e_{2,1} \quad , \quad h = e_{1,1} - e_{2,2}$$

and

$$\mathfrak{n}_+ = \mathbf{C} \cdot x \quad , \quad \mathfrak{n}_- = \mathbf{C} \cdot y \quad , \quad \mathfrak{h} = \mathbf{C} \cdot h$$

The Cartan subalgebra is one-dimensional, in this case. Subsequently we write the simple Lie algebra $\mathfrak{sl}_2$ as

$$\mathfrak{sl}_2 = \mathfrak{n}_+ \oplus \mathfrak{h} \oplus \mathfrak{n}_-$$

Then define

$$\hat{\mathfrak{n}}_+ = \mathfrak{n}_+ \oplus (\mathfrak{sl}_2 \otimes t\mathbf{C}[t])$$

and

$$\hat{\mathfrak{n}}_- = \mathfrak{n}_- \oplus (\mathfrak{sl}_2 \otimes t^{-1}\mathbf{C}[t^{-1}])$$

where $\mathbf{C}[t]$ and $\mathbf{C}[t^{-1}]$ denote polynomials in t and t^{-1}, respectively, $t\mathbf{C}[t]$ and $t^{-1}\mathbf{C}[t^{-1}]$ those with vanishing constant term, respectively. Thus, $\tau(a)$ is a strictly upper triangular matrix (in $\mathfrak{a}'_\infty$) if and only if $a \in \hat{\mathfrak{n}}_+$, where τ is the injective homomorphism of $\widetilde{\mathfrak{gl}}_2$ into $\mathfrak{a}'_\infty$, discussed earlier.

The Cartan subalgebra $\hat{\mathfrak{h}}$ of $\widehat{\mathfrak{sl}}_2$ is spanned by $\hat{h}_1 = e_{1,1}(0) - e_{2,2}(0)$, c and d, however, we will usually prefer the basis $\{\hat{h}_0 = c - \hat{h}_1, \hat{h}_1, d\}$ for $\hat{\mathfrak{h}}$. We may write

$$\widehat{\mathfrak{sl}}_2 = \hat{\mathfrak{n}}_+ \oplus \hat{\mathfrak{h}} \oplus \hat{\mathfrak{n}}_-$$

The linear functionals ω_0 and ω_1 on $\mathfrak{h}$ given by

$$\omega_j(\hat{h}_i) = \delta_{i-j}$$

and

$$\omega_j(d) = 0$$

for $i, j = 0, 1$, are usual called the fundamental weights. Observe that

$$
\begin{aligned}
\pi_q(\tau(\hat{h}_1))\Omega_q &= \left(\pi_q(\tau(e_{1,1}(0))) - \pi_q(\tau(e_{2,2}(0)))\right)\Omega_q \\
&= \left(\sum_{k \in \mathbf{Z}} \lambda_{2k+1} - \sum_{k \in \mathbf{Z}} \lambda_{2k+2}\right)\Omega_q \\
&= \omega_{(q \bmod 2)}(\hat{h}_1)\Omega_q \\
&= \omega_p(\hat{h}_1)\Omega_q
\end{aligned}
$$

where $p = p(q) = q \bmod 2 \in \{0, 1\}$ and

$$
\begin{aligned}
\pi_q(\tau(\hat{h}_0))\Omega_q &= \left(\pi_q(c) - \pi_q(\tau(\hat{h}_1))\right)\Omega_q \\
&= \left(1 - \omega_p(\hat{h}_1)\right)\Omega_q \\
&= \omega_p(\hat{h}_0)\Omega_q
\end{aligned}
$$

That is

$$\pi_q(\tau(\hat{h}_i)) = \omega_p(\hat{h}_i)\Omega_q$$

where $p = q \bmod 2$. If we put $\pi_q(\tau(d))\Omega_q = 0$ (as later shown is allowed) then $\pi_q(\tau(\hat{h}))\Omega_q = \omega_p(\hat{h})\Omega_q$ for all $\hat{h} \in \mathfrak{h}$.

Following [K-R, p 101] we define $\lambda \in \hat{\mathfrak{h}}^*$ to be a highest weight and π_λ a highest weight representation of the Lie algebra $\hat{\mathfrak{sl}}_2$ if the representation space $\mathcal{L}(\lambda)$ contains a non-zero vector ν_λ such that π_λ is irreducible and fulfils

$$\pi_\lambda(\hat{n}_+)\nu_\lambda = 0$$

and

$$\pi_\lambda(\hat{h})\nu_\lambda = \lambda(\hat{h})\nu_\lambda$$

for $\hat{h} \in \hat{\mathfrak{h}}$. The distinguished vector ν_λ is called the highest weight vector.

It immediately follows that for $q \in \mathbf{Z}$ the above constructed representation $\pi_q(\tau(\cdot))$ of $\hat{\mathfrak{sl}}_2$ fulfils that $\pi_q(\tau(\hat{n}_+))\Omega_q = 0$ and $\pi_q(\tau(\hat{h}))\Omega_q = \omega_p(\hat{h})\Omega_q$, where $p = q \bmod 2$. So we are rather close to having constructed a highest weight representation of $\widehat{\mathfrak{sl}}_2$ with Ω_q as the highest weight vector, but $\pi_q(\tau(\cdot))$ is not irreducible. However, each $\mathcal{H}_q$ does have a proper invariant subspace $\mathcal{H}'_q$ containing Ω_q, though this does not in itself ensure any proper invariant subspace containing Ω_q. Hence, we arrive with a highest weight representation of $\widehat{\mathfrak{sl}}_2$.

Recall that $\pi_q(\tau(\cdot))$ of $\widehat{\mathfrak{gl}}'_2$ in $\mathcal{H}_q$ is irreducible, since $\tau(\widehat{\mathfrak{gl}}'_2)$ contains the shift operators s_k, $k \in \mathbf{Z}$. As shown earlier $s_{2k} \notin \tau(\widehat{\mathfrak{sl}}_2)$ but $s_{2k+1} \in \tau(\widehat{\mathfrak{sl}}'_2)$, which leads us to define

$$\mathcal{H}'_q = \{F \in \mathcal{H}_q : \pi_q(s_{2k})F = 0, k \in \mathbf{N}\}$$

It follows that $\mathcal{H}'_q$ is invariant under $\pi_q(\tau(\widehat{\mathfrak{sl}}'_2))$, since $[s_{2k}, \tau(\widehat{\mathfrak{sl}}'_2)] = 0$, as shown earlier. Observe that

$$\pi_q(s_{-2k_1+1}) \cdot \ldots \cdot \pi_q(s_{-2k_n+1})\Omega_q$$

with $k_1 \geq \cdots \geq k_n > 0$, span a dense set in $\mathcal{H}'_q$, since

$$\pi_q(s_{2k})\pi_q(s_{-j_1}) \cdot \ldots \cdot \pi_q(s_{-j_n})\Omega_q$$
$$= \pi_q(s_{-j_1}) \cdot \ldots \cdot \pi_q(s_{-j_n})\pi_q(s_{2k})\Omega_q$$
$$+ \sum_{i=1}^{n} \delta_{2k-j_i}\, 2k\, \pi_q(s_{-j_1}) \cdot \ldots \cdot \pi_q(s_{-j_{i-1}})\pi_q(s_{-j_{i+1}}) \cdot \ldots \cdot \pi_q(s_{-j_n})\Omega_q$$

Thus we have the following theorem.

Theorem 31 *For each $q \in \mathbf{Z}$, π_q is a unitary highest weight representation of $\widehat{\mathfrak{sl}}_2$ with highest weight ω_p, $p = q \bmod 2$, in $\mathcal{H}'_q$, and highest weight vector Ω_q.*

Proof. An immediately consequence of the above and the fact that the inner product defines a contravariant positive definite Hermitian form with respect to the anti-linear anti-involution given by the restriction of that on $\widehat{\mathfrak{gl}}_2$ and by $\omega(c) = c$ combined with $\omega(d) = d$. $\square$

Due to the fact that the highest weight in the theorem 31 is a fundamental weight, we call the corresponding highest weight representations fundamental representations. Since in the present case there are two possible different fundamental weights, ω_0 and ω_1, it follows that there exists only two essentially different fundamental representations $\pi_0(\tau(\cdot))$ and $\pi_1(\tau(\cdot))$, because if $p = q \bmod 2$ then $\omega_{(p \bmod 2)} = \omega_{(q \bmod 2)}$ and the mapping given by $\Omega_q \to \Omega_p$ and

$$\pi_q(s_{-2k_1+1}) \cdot \ldots \cdot \pi_q(s_{-2k_n+1})\Omega_q \to \pi_p(s_{-2k_1+1}) \cdot \ldots \cdot \pi_p(s_{-2k_n+1})\Omega_p$$

extends by linearity and continuity to a unitary operator U from $\mathcal{H}'_q$ onto $\mathcal{H}'_p$, so $\pi_p(\tau(\cdot))$ and $\pi_q(\tau(\cdot))$ become unitarily equivalent, $\pi_p(\tau(\cdot)) = U\pi_q(\tau(\cdot))U^*$.

Corollary 32 *The representations* $\pi_\lambda = \pi_0(\tau(\cdot))^{\otimes k_0} \otimes \pi_1(\tau(\cdot))^{\otimes k_1}$, *of* $\widehat{\mathfrak{sl}}_2$ *in* $\mathcal{L}(\lambda) = \pi_\lambda(\widehat{\mathfrak{sl}}_2)(\Omega_0^{\otimes k_0} \otimes \Omega_1^{\otimes k_1})$, *where* $\lambda = k_0\omega_0 + k_1\omega_1$ *and* $(k_0, k_1) \in (\mathbf{N} \cup \{0\})^2 \setminus \{(0,0)\}$ *are unitary highest weight representations with highest weight* λ *and highest weight vector* $\Omega_0^{\otimes k_0} \otimes \Omega_1^{\otimes k_1}$.

Proof. As mentioned earlier, the tensor product of unitary highest weight representations (which are irreducible by definition) defines a unitary highest weight representation in the highest component, i.e. the vector space generated by the tensor product of the highest weight vectors, with highest weight given by the sum of the highest weights corresponding to the involved representations. □

The converse of the above corollary is included in the following proposition.

Proposition 33 *The highest weight representations* π_λ *of* $\widehat{\mathfrak{sl}}_2$ *in* $\mathcal{L}(\lambda)$ *are unitary if and only if* $\lambda(\hat{h}_i) \in \mathbf{N}$, *for* $i = 0, 1$, *and* $\lambda(d) \in \mathbf{R}$.

Proof. The if part follows directly from the above corollary, since we may subtract $\lambda(d)$ from d giving $d' = d - \lambda(d)$, thus $\lambda(d') = 0$.

The only if part follows since we may identify $\mathfrak{sl}_2$ naturally with the subalgebra of $\mathfrak{sl}_2$ generated by

$$x(0) = e_{1,2}(0) \quad , \quad y(0) = e_{2,1}(0) \quad , \quad h(0) = \hat{h}_1 = e_{1,1}(0) - e_{2,2}(0)$$

The restriction of the unitary representation π_λ to $\mathfrak{sl}_2$ is still unitary. Moreover, it becomes irreducible in $\{\pi_\lambda(y(0))^k \nu_\lambda : k \in \mathbf{N} \cup \{0\}\}$, where ν_λ denotes the highest weight vector, because $\pi_\lambda(x(0))\nu_\lambda = 0$ and $\pi_\lambda(h(0))\nu_\lambda = \lambda(h(0))\nu_\lambda$. Thus, π_λ is an irreducible unitary highest weight representation in the above space. In the following we suppress the representation symbol π_λ. Then the action of $\mathfrak{sl}_2$ in

$\{y(0)^k \nu_\lambda : k \in \mathbf{N} \cup \{0\}\}$ is irreducible and unitary. Direct calculation using the commutation relations gives

$$x(0)y(0)^k \nu_\lambda = \sum_{i=0}^{k-1} y(0)^i [x(0), y(0)] y(0)^{k-i-1} \nu_\lambda + y(0)^k x(0) \nu_\lambda$$

$$= \sum_{i=0}^{k-1} y(0)^i h(0) y(0)^{k-i-1} \nu_\lambda$$

and

$$h(0)y(0)^n \nu_\lambda = y(0)^n \left(h(0) - 2n \right) \nu_\lambda$$
$$= y(0)^n \left(\lambda(h(0)) - 2n \right) \nu_\lambda$$
$$= \left(\lambda(h(0)) - 2n \right) y(0)^n \nu_\lambda$$

by repeated use of

$$h(0)y(0) = [h(0), y(0)] + y(0)h(0)$$
$$= -2y(0) + y(0)h(0)$$
$$= y(0) \left(h(0) - 2 \right)$$

Hence

$$x(0)y(0)^k \nu_\lambda = \sum_{i=0}^{k-1} y(0)^i \left(\lambda(h(0)) - 2(k - i - 1) \right) y(0)^{k-i-1} \nu_\lambda$$

$$= \sum_{i=0}^{k-1} \left(\lambda(h(0)) - 2(k - i - 1) \right) \left(y(0)^{k-1} \nu_\lambda \right)$$

$$= \left(k\lambda(h(0)) - 2\frac{1}{2}k(k - 1) \right) \left(y(0)^{k-1} \nu_\lambda \right)$$

$$= k \left(\lambda(h(0)) - k + 1 \right) \left(y(0)^{k-1} \nu_\lambda \right)$$

Then for $\lambda(h(0)) \notin \mathbf{N} \cup \{0\}$, $y(0)^k \nu_\lambda \neq 0$, by iteration, and because $y(0)^k \nu_\lambda$, $k \in \mathbf{N} \cup \{0\}$, are linearly independent, since different vectors corresponds to different eigenvalues of $h(0)$ by the above calculation, it follows that $\{y(0)_k \nu_\lambda : k \in \mathbf{N} \cup \{0\}\}$ span an infinite-dimensional space. But the only unitary irreducible representations of $\mathfrak{sl}_2$ with the given involution are the finite dimensional ones. Since the compact real form is the real subspace $\{a \in \mathfrak{sl}_2 : a^* = -a\}$, which, in fact, equals $\mathfrak{su}(2,\mathbf{R})$, and the unitary irreducible representation restricted to $\mathfrak{su}(2,\mathbf{R})$ exponentiate to a unitary irreducible group representation of the corresponding Lie group $Su(2,\mathbf{R})$, the claim follows, because all unitary irreducible representations of $Su(2,\mathbf{R})$ are known and are finite dimensional, since $Su(2,\mathbf{R})$ is a compact group (see [Dix, corollary 15.1.4]). Thus, $\lambda(h(0)) \in \mathbf{N} \cup \{0\}$ and then also $\lambda(\hat{h}_1) \in \mathbf{N} \cup \{0\}$, since $\hat{h}_1 = h(0)$. We emphasize that if $\lambda(h(0)) = n - 1 \in \mathbf{N} \cup \{0\}$ then the representation space becomes $\{y(0)^k \nu_\lambda : k = 0, 1, \ldots, n - 1\}$, which has dimension $n \in \mathbf{N}$.

There is another copy of $\mathfrak{sl}_2$ in $\widehat{\mathfrak{sl}}_2$, different from the one considered above, namely the Lie algebra spanned by the elements

$$x_1 = e_{2,1}(1) \quad , \quad y_1 = e_{1,2}(-1) \quad , \quad h_1 = [x_1, y_1] = c - \hat{h}_1 = \hat{h}_0$$

since

$$[h_1, x_1] = 2x_1 \qquad \text{and} \qquad [h_1, y_1] = -2y_1$$

by direct calculation using the earlier derived commutator relations (on $\widehat{\mathfrak{gl}}_2'$) $[a(k), b(l)] = [a, b]_0(k + l) + k\delta_{k+l}\mathrm{Tr}(ab) \cdot c$ and that $[x_1, y_1] = h_1$ by definition. Now, by considerations as above it follows that $\lambda(h_1) = \lambda(\hat{h}_0) \in \mathbf{N} \cup \{0\}$. Then $\lambda(c) = \lambda(h_1) + \lambda(h(0)) \in \mathbf{N} \cup \{0\}$, since $\hat{h}_1 = h(0)$. Hereby the proposition is proved.   □

Since unitarity of an irreducible representation of $\widehat{\mathfrak{sl}}_2$ implies $\lambda(c) \in \mathbf{N} \cup \{0\}$ and that this distinguished value somehow characterizes the representation (see below) it is sometimes referred to as *the level* of the particular representation.

We will now demonstrate how the Goddard-Kent-Olive (GKO) construction gives a recipe to construct a series of unitary representations of the Virasoro algebra with central charge running through a discrete subset of $[0, 1[$. The GKO construction uses the Sugawara construction, described in the last part of section 4.1, together with theorem 26 of section 4.2. To obtain the mentioned series of representations, the GKO construction will be built on $\widehat{\mathfrak{sl}}_2'$ and we therefore limit ourselves to this case in the following considerations. However, the GKO construction does have meaning when built on an arbitrary simple Lie algebra.

Put $\mathfrak{g} = \mathfrak{sl}_2 \oplus \mathfrak{sl}_2$ and consider two representations of $\widehat{\mathfrak{sl}}_2$ in vector spaces $\mathcal{L}(\lambda) = \mathcal{H}'_q$ for some $q \in \mathbf{Z}$ and $\mathcal{L}(\mu) = \mathcal{H}'_p$ for some $p \in \mathbf{Z}$ with levels $m_\lambda = \lambda(c)$ and $m_\mu = \mu(c)$, respectively. In what follows we suppress the representation symbols, as previously done. Then the action of $\hat{\mathfrak{g}}' = \widehat{\mathfrak{sl}_2}' \oplus \widehat{\mathfrak{sl}_2}'$ in $\mathcal{L}(\lambda) \otimes \mathcal{L}(\mu)$ is given by

$$(x(n) \oplus y(m))(v \otimes w) = (x(n)(v)) \otimes w + v \otimes (y(m)(w))$$

for any $v \otimes w \in \mathcal{L}(\lambda) \otimes \mathcal{L}(\mu)$ and $n, m \in \mathbf{Z}$. Thus, the Sugawara construction (see section 4.1) of the representations $x(n) \otimes 1$ and $1 \otimes y(m)$ provides us with Virasoro operators $L_k^{(\lambda)} \otimes 1$ and $1 \otimes L_k^{(\mu)}$, respectively. Hence

$$
\begin{aligned}
L_k^{\mathfrak{g}} &= L_k^{(\lambda)} \otimes 1 + 1 \otimes L_k^{(\mu)} \\
&= \frac{1}{2(Q + m_\lambda)} \sum_{l \in \mathbf{Z}} \sum_{i=1}^{3} (:x_i(l)x_i(k-l):) \otimes 1 \\
&\quad + \frac{1}{2(Q + m_\mu)} \sum_{l \in \mathbf{Z}} \sum_{i=1}^{3} 1 \otimes (:y_i(l)y_i(k-l):)
\end{aligned}
$$

defines a representation of the Virasoro algebra in $\mathcal{L}(\lambda) \otimes \mathcal{L}(\mu)$, with central charge

$$c_{\mathfrak{g}} = c_\lambda + c_\mu = 3 \left(\frac{m_\lambda}{Q + m_\lambda} + \frac{m_\mu}{Q + m_\mu} \right)$$

since

$$
\begin{aligned}
[L_n^{\mathfrak{g}}, L_k^{\mathfrak{g}}] &= (n-k)L_{n+k}^{(\lambda)} \otimes 1 + \delta_{n+k} \frac{n^3 - n}{12} \cdot \frac{m_\lambda \dim(\mathfrak{sl}_2)}{Q + m_\lambda} \\
&\quad + (n-k)1 \otimes L_{n+k}^{(\mu)} + \delta_{n+k} \frac{n^3 - n}{12} \cdot \frac{m_\mu \dim(\mathfrak{sl}_2)}{Q + m_\mu} \\
&= (n-k)L_{n+k}^{\mathfrak{g}} + \delta_{n+k} \frac{n^3 - n}{12} \cdot c_{\mathfrak{g}}
\end{aligned}
$$

and the dimension of $\mathfrak{sl}_2$ is 3. However, there is also a level $m_\lambda + m_\mu$ representation of $\widehat{\mathfrak{sl}_2}'$ ($\cong \{x \otimes y \in \widehat{\mathfrak{sl}_2} \otimes \widehat{\mathfrak{sl}_2} : x = y\}$) in $\mathcal{L}(\lambda) \otimes \mathcal{L}(\mu)$, namely the restriction of the former to the diagonal (or $x(n) \to x(n) \otimes 1 + 1 \otimes x(n)$). By the Sugawara construction of this representation, we obtain the following Virasoro operators

$$
\begin{aligned}
L_k' &= \frac{1}{2(Q + m_\lambda + m_\mu)} \sum_{l \in \mathbf{Z}} \sum_{i=1}^{3} :(x_i(l) \otimes 1 + 1 \otimes x_i(l)) \\
&\qquad\qquad\qquad \cdot (x_i(k-l) \otimes 1 + 1 \otimes x_i(k-l)): \\
&= \frac{1}{2(Q + m_\lambda + m_\mu)} \sum_{l \in \mathbf{Z}} \sum_{i=1}^{3} ((:x_i(l)x_i(k-l):) \otimes 1 \\
&\qquad + 1 \otimes (:x_i(l)x_i(k-l):) + 2x_i(l) \otimes x_i(k-l))
\end{aligned}
$$

with central charge $c' = 3\frac{m_\lambda + m_\mu}{Q + m_\lambda + m_\mu}$, since $\dim(\mathfrak{sl}_2) = 3$. Thus, theorem 26 gives us the Virasoro operators

$$L_k = L_k^\mathfrak{g} - L_k'$$

$$= \left(\frac{1}{2(Q + m_\lambda)} - \frac{1}{2(Q + m_\lambda + m_\mu)} \right) \sum_{l \in \mathbf{Z}} \sum_{i=1}^{3} (:x_i(l)x_i(k-l):) \otimes 1$$

$$+ \left(\frac{1}{2(Q + m_\mu)} - \frac{1}{2(Q + m_\lambda + m_\mu)} \right) \sum_{l \in \mathbf{Z}} \sum_{i=1}^{3} 1 \otimes (:x_i(l)x_i(k-l):)$$

$$- \frac{1}{Q + m_\lambda + m_\mu} \sum_{l \in \mathbf{Z}} \sum_{i=1}^{3} x_i(l) \otimes x_i(k-l)$$

Hereby we get a representation of the Virasoro algebra, in $\mathcal{L}(\lambda) \otimes \mathcal{L}(\mu)$, with central charge

$$c = c_\mathfrak{g} - c' = 3 \left(\frac{m_\lambda}{Q + m_\lambda} + \frac{m_\mu}{Q + m_\mu} - \frac{m_\lambda + m_\mu}{Q + m_\lambda + m_\mu} \right)$$

One can calculate L_0 explicetly. Notice that $\Omega_\lambda = 2(Q + m_\lambda)d + 2T(0)$ commutes with every element of $\widehat{\mathfrak{sl}}_2$, where

$$T(0) = \frac{1}{2} \sum_{i=1}^{3} x_i(0)x_i(0) + \sum_{l \in \mathbf{N}} \sum_{i=1}^{3} x_i(-l)x_i(l)$$

see section 4.1, and Ω_λ is the Casimir operator of $\widehat{\mathfrak{sl}}_2$ in $\mathcal{L}(\lambda)$ since

$$[d, \Omega_\lambda] = 0 + 2[d, T(0)] = 0$$

and

$$[x(n), \Omega_\lambda] = -2(Q + m_\lambda)n\, x(n) + 2(Q + m_\lambda)n\, x(n) = 0$$

by direct calculation using the various commutation relations on $\widehat{\mathfrak{sl}}_2$ given in section 4.1 together with lemma 24. Trivially, $[c_\lambda, \Omega_\lambda] = 0$. We have the similar relations with μ instead of λ and also for the diagonal representation. Then

$$L_0^{(\lambda)} = \frac{1}{Q + m_\lambda} T(0) = \frac{\Omega_\lambda}{2(Q + m_\lambda)} - d,$$

$$L_0^{(\mu)} = \frac{\Omega_\mu}{2(Q + m_\mu)} - d$$

and

$$L_0' = \frac{\Omega'}{2(Q + m_\lambda + m_\mu)} - d \otimes 1 - 1 \otimes d$$

where Ω' is the Casimir operator of $\widehat{\mathfrak{sl}}_2$ in $\mathcal{L}(\lambda) \otimes \mathcal{L}(\mu)$, whenever we are considering admissible representations, as the highest weight representations indeed are. We consequently obtain the following expression

$$\begin{aligned} L_0 &= \left(\frac{\Omega_\lambda}{2(Q + m_\lambda)} - d\right) \otimes 1 + 1 \otimes \left(\frac{\Omega_\mu}{2(Q + m_\mu)} - d\right) \\ &\quad - \left(\frac{\Omega'}{2(Q + m_\lambda + m_\mu)} - d \otimes 1 - 1 \otimes d\right) \\ &= \frac{1}{2}\left(\frac{(\lambda, \lambda + 2\rho)}{Q + m_\lambda} + \frac{(\mu, \mu + 2\rho)}{Q + m_\mu} - \frac{\Omega'}{Q + m_\lambda + m_\mu}\right) \end{aligned}$$

where we have used that the eigenvalue of Ω_ν is $(\nu, \nu + 2\rho)$, for $\nu = \lambda, \mu$, and ρ is the sum of fundamental weights (see [Hu, section 22.3] or [K-R, section 10.1-10.2]). However, we will not really need the expression for L_0, but rather focus on the possible values of c. We therefore return to the Cartan subalgebra $\hat{\mathfrak{h}}$ of $\widehat{\mathfrak{sl}}_2$. Choose $\{b = \hat{h}_1 = e_{1,1}(0) - e_{2,2}(0), c, d\}$ as a basis for $\hat{\mathfrak{h}}$, hence $\hat{\mathfrak{h}} = \mathbf{C} \cdot b \oplus \mathbf{C} \cdot c \oplus \mathbf{C} \cdot d$. The bilinear form on $\widehat{\mathfrak{gl}}_2$ (from proposition 20, with $s = 0$ and $(a(t), b(t))_0 = \mathrm{Tr}(a(t)b(t))$, $a(t), b(t) \in \widehat{\mathfrak{gl}}_2$) restricts to a bilinear form on $\widehat{\mathfrak{sl}}_2$ and becomes non-degenerated when restricted further to $\hat{\mathfrak{h}}$. In fact

$$\begin{aligned} (b, b) &= \mathrm{Tr}\left((e_{1,1}(0) - e_{2,2}(0))(e_{1,1}(0) - e_{2,2}(0))\right) \\ &= \mathrm{Tr}\left(e_{1,1}(0) + e_{2,2}(0)\right) = 2 \end{aligned}$$

and

$$(c, d) = (d, c) = 1$$

All other evaluations of the form vanish. We may then identify $\hat{\mathfrak{h}}$ with the dual space $\hat{\mathfrak{h}}^*$ by this form. In the highest weight representation $\mathcal{L}(\lambda)$ of $\widehat{\mathfrak{sl}}_2$ the action of $\hat{\mathfrak{h}}$ on the highest weight vector ν_λ is given as

$$\hat{h}(\nu_\lambda) = \lambda(\hat{h})\nu_\lambda = (\lambda, \hat{h})\nu_\lambda$$

for $\hat{h} \in \hat{\mathfrak{h}}$. With the above choice of basis the fundamental weights become $\omega_0 = (d, \cdot) \cong d$ and $\omega_1 = (\frac{1}{2}b, \cdot) \cong \frac{1}{2}b$, since $\omega_i(\hat{h}_j) = \delta_{i-j}$,

$i = 0, 1$, $\hat{h}_0 = c$ and $\hat{h}_1 = b$. As earlier ρ denotes the sum of the fundamental weights, $\rho = \omega_0 + \omega_1 \cong d + \frac{1}{2}b$.

We know from proposition 33 that the representation in $\mathcal{L}(\lambda)$ is unitary if and only if λ takes the form

$$\lambda = m \cdot d + \frac{1}{2}n \cdot b + r \cdot c$$

where $r = \lambda(d) \in \mathbf{R}$, $m = \lambda(c) \in \mathbf{N} \cup \{0\}$ and $n = \lambda(b) \in \mathbf{N} \cup \{0\}$. Then $c(\nu_\lambda) = \lambda(c)\nu_\lambda = m \cdot \nu_\lambda$ and $\mathcal{L}(\lambda)$ becomes a level m representation $(m_\lambda = m)$. Choosing the standard basis $\{x, y, h\}$ for $\mathfrak{sl}_2$, it follows directly that $x' = y$, $y' = x$ and $h' = \frac{1}{2}h$ defines the dual basis with respect to the trace form, $(a, b)_0 = \mathrm{Tr}(ab)$, restricted from $\mathfrak{gl}_2$.

By use of the commutation relations on $\mathfrak{sl}_2$ it follows that the Casimir operator becomes

$$\Omega_0 = xx' + hh' + yy' = xy + \frac{1}{2}h^2 + yx = \frac{1}{2}h^2 + h + 2yx$$

and in the adjoint representation

$$\Omega_0^{(\mathrm{ad})} = (\mathrm{ad}x)(\mathrm{ad}y) + \frac{1}{2}(\mathrm{ad}h)(\mathrm{ad}h) + (\mathrm{ad}y)(\mathrm{ad}x)$$

which has eigenvalue $2Q = 4$, since it is known that it acts as a scalar $2Q$ and it follows directly that $(\Omega_0^{(\mathrm{ad})})(h) = 4h$. Hence, $Q = 2$. We now choose some particular weights $\mu = d$ (corresponding to $m = 1$, $n = 0$ and $r = 0$) and $\lambda = md + \frac{1}{2}nb$ (corresponding to $r = 0$) with $m, n \in \mathbf{N} \cup \{0\}$. Thus, the previously introduced GKO construction for $\mathfrak{sl}_2$ provides us with unitary representations of the Virasoro algebra, with $Q = 2$ and $m_\mu = \mu(c) = (d, c) = 1$. It becomes a highest weight representation in the highest component. The operator L_0, dependent on m and n, becomes

$$\begin{aligned} L_0 &= \frac{1}{2}\left(\frac{(md + \frac{1}{2}nb, md + \frac{1}{2}nb + 2d + b)}{2 + m} \right. \\ &\qquad \left. + \frac{(d, d + 2d + b)}{2 + 1} - \frac{\Omega'}{2 + m + 1} \right) \\ &= \frac{\frac{1}{2}n^2 + n}{4(m + 2)} + 0 - \frac{\Omega'}{2(m + 3)} = \frac{n(n + 2)}{4(m + 2)} - \frac{\Omega'}{2(m + 3)} \end{aligned}$$

where $m, n \in \mathbf{N} \cup \{0\}$. The interesting central charge in this particular case becomes

$$c_m = 3 \left(\frac{m}{2+m} + \frac{1}{2+1} - \frac{m+1}{2+m+1} \right) = 1 - \frac{6}{(m+2)(m+3)}$$

where $m \in \mathbf{N} \cup \{0\}$.

Hereby we have succeded in constructing a series of unitary highest weight representations of the Virasoro algebra in the highest
component of $\mathcal{L}(\lambda) \otimes \mathcal{L}(\mu) = \mathcal{H}'_q \otimes \mathcal{H}'_p$ generated by $v_\lambda \otimes v_\mu (= \Omega_q \otimes \Omega_p)$,
with central charge given by the above discrete series $c_m \in [0, 1[$. We
emphasize that $c_1 = \frac{1}{2}$ for $m = 1$.

It should be pointed out that the eigenvalue $h \in \mathbf{R}$ of L_0 also
belongs to a discrete series (here h doesn't denote an element of $\mathfrak{h}$
or $\hat{\mathfrak{h}}$). In fact, for a given $m \in \mathbf{N} \cup \{0\}$ or, equivalent, a given c_m,
the $\frac{1}{2}(m+1)(m+2)$ numbers

$$h_m(r, s) = \frac{((m+3)r - (m+2)s)^2 - 1}{4(m+2)(m+3)}$$

where $r, s \in \mathbf{N}$ such that $1 \leq s \leq r \leq m+1$, are the only possible
values h can take. We will not dwell on the proof of this, but mention
that it is based on the celebrated Kac determinant formula (see for
example [K-R, Chap. 8-12]). From this formula it follows that for
$m = 1$ or, equivalently, $c = \frac{1}{2}$, the only possible values of h giving
unitary representations are those where

$$h = h_1(1, 1) = 0 \quad , \quad h = h_1(1, 2) = \frac{1}{2} \quad , \quad h = h_1(2, 2) = \frac{1}{16}$$

a result which we have used earlier in section 4.2.

Chapter 5

Applications

5.1 The Loop Group LS^1

In this section we consider the particular loop group LS^1 viewed as an abelian subgroup of the restricted unitary group, discussed in section 2.3. We will, in particular, use the infinite-dimensional spin representation, treated in detail in sections 2.4 and 2.3, in our discussion of the particular loop group LS^1, also known as the loop circle. Section 4.1 on loop algebras will serve as background material.

The loop groups have been studied intensively in the book by Pressley and Segal [P-S] from 1986. Even though their approach is different from ours [P-S] will serve as a foundation of this section. The loop circle has also been considered briefly by Lundberg in [Lu 2], we will follow and elaborate this approach. Several others have studied the loop groups (see [P-S] for further references), among them [Mi].

The applications of loop groups are various, for example loop groups appear in two dimensional quantum field theory and more recently they have been put to extensive use in connection with the so-called completely integrable systems of partial differential equations (see [P-S] and [Mi] for further details).

Consider the complex Hilbert space $\mathcal{H} = L_2(S^1)$ of square-integrable functions on the unit circle S^1. The inner product on $\mathcal{H}$ is given by

$$\langle f, g \rangle = \frac{1}{2\pi} \int_0^{2\pi} \overline{f(\theta)} g(\theta) \, d\theta$$

for any $f, g \in \mathcal{H}$.

We may choose $\{e_k \in \mathcal{H} : k \in \mathbf{Z}\}, e_k(\theta) = e^{ik\theta}$, as an orthonormal basis for $\mathcal{H}$. We then have a canonical splitting, or polarization, of $\mathcal{H}$ as $\mathcal{H} = \mathcal{H}_+ \oplus \mathcal{H}_-$, where $\mathcal{H}_+ = \mathrm{span}\{e_k : k \geq 0\}$ and $\mathcal{H}_- = \mathrm{span}\{e_k : k < 0\}$. We denote the projection onto $\mathcal{H}_-$ by P.

The loop group LG is the group of parameterized loops in the group G, i.e. LG is the group of smooth maps (loops) from the circle S^1 into G. In the present case we only consider $G = S^1$. Hence, LS^1 consists of smooth endomorphisms of S^1. Then any element in LS^1 can be written in the form e^{iF}, where F is a smooth function from S^1 into $\mathbf{R}$ such that $F(\theta + 2\pi) = F(\theta) + 2\pi \cdot n_F$, for some integer n_F, which is the winding number of e^{iF} (see for example [P-S, p. 59]). Notice that $e^{iF} : S^1 \to S^1$, $e^{iF(\theta+2\pi)} = e^{iF(\theta)}$ and e^{iF} is indeed smooth.

Due to [P-S, p. 79] we may realize the *loop group LS^1* as multiplication operators of the form $M(e^{iF})$ on $\mathcal{H}$, i.e. $(M(e^{iF})g)(\theta) = e^{iF(\theta)} \cdot g(\theta)$, for any $g \in \mathcal{H}$. In the following we simply write e^{iF} for $M(e^{iF})$, whereby we identify the multiplication operator $M(e^{iF})$ on $\mathcal{H}$ with the group element $e^{iF} \in LS^1$. In the present case it is sufficient to demand that the real valued function F on S^1 is C^1, i.e. $F \in C^1(S^1)$, and not necessarily smooth. If we put $f_F(\theta) = F(\theta) - n_F \cdot \theta$ then $f_F(\theta + 2\pi) = f_F(\theta)$. Hence $f_F \in C^1(S^1)$ is real and invariant with respect to the shift $\theta \to \theta + 2\pi$. Therefore f_F can be expanded as a Fourier series $f_F = \frac{1}{2\pi} \sum_{k \in \mathbf{Z}} f_k \cdot e_k$, where the Fourier coefficients f_k are given by $f_k = \langle e_k, f_F \rangle = \frac{1}{2\pi} \int_0^{2\pi} f_F(\theta) e^{-ik\theta} \, d\theta$. Separating out the zero'th Fourier component $f_0 = \frac{1}{2\pi} \int_0^{2\pi} f_F(\theta) \, d\theta \in \mathbf{R}$ we may write $f_F = f_0 + \frac{1}{2\pi} \sum_{k \in \mathbf{Z} \setminus \{0\}} f_k \cdot e_k$, where $e_k(\theta) = e^{ik\theta}$ and $\overline{f}_{-k} = f_k$ (since f_F is real). Put $f(\theta) = \frac{1}{2\pi} \sum_{k \in \mathbf{Z} \setminus \{0\}} f_k \cdot e_k(\theta) = f_F(\theta) - f_0$, and notice that the mean value of f is $\frac{1}{2\pi} \int_0^{2\pi} f(\theta) \, d\theta = 0$. We finally come up with the following splitting of F

$$F(\theta) = n_F \cdot \theta + f_0 + f(\theta)$$

where $n_F \in \mathbf{Z}$ is the winding number of e^{iF}, $f_0 \in \mathbf{R}$ is the mean value of $F(\theta) - n_F \cdot \theta$ and $f \in C^1(S^1)$ is real and has zero-trace, $\int_0^{2\pi} f(\theta) \, d\theta = 0$. If we put $F_0(\theta) = n_F \cdot \theta + f_0$, such that $F = F_0 + f$, then $e^{iF} \in LS^1$ can be factorized as $e^{iF_0} \cdot e^{if}$. The subgroup of LS^1 consisting of elements of the form e^{if} will be called *the special loop group*, denoted by SLS^1, and the subgroup of LS^1 generated by elements of the form e^{iF_0} will be called *the charge subgroup of LS^1*, denoted by C. Thus LS^1 may be considered as the product group $SLS^1 \times C$.

We consider the case of the special loop group SLS^1 first. Below, we show that SLS^1 may be considered as a subgroup of the restricted unitary group $\mathcal{U}_2(\mathcal{H}, P)$, studied in section 2.3, where P is the projection onto $\mathcal{H}_-$. Furthermore, we study the explicit spin representation, which is a projective representation of positive energy of SLS^1.

According to the results in section 2.3, we only have to verify that $if \in u_2(\mathcal{H}, P)$, whence $e^{if} \in \mathcal{U}_2(\mathcal{H}, P)$. Evidently $(if)^* = -if$. The Hilbert-Schmidt stipulation of $[P, if]$ is equivalent to that of $Pf(1 - P) = P[P, f]$, since $[P, f] = Pf(1 - P) - (1 - P)fP$ and $Pf(1-P)$ is Hilbert-Schmidt if and only if $(1-P)fP = (Pf(1-P))^*$ is. Therefore it is sufficient to prove that $Pf(1 - P)$ is Hilbert-Schmidt. Let χ denote the indicator function of $\mathbb{N} \cup \{0\}$, i.e. $\chi(k) = 1$ for $k \in \mathbb{N} \cup \{0\}$ and zero otherwise. Then

$$\langle e_k, Pf(I - P)fPe_k \rangle = \sum_{n \in \mathbb{Z}} \chi(-k - 1)\chi(n + k) \cdot |f_n|^2$$

by direct calculation using the Fourier series for f (and the fact that $\overline{f}_{-n} = f_n$, since f is real). Hence

$$
\begin{aligned}
\mathrm{Tr}(Pf(I - P)fP) &= \sum_{k \in \mathbb{Z}} \langle e_k, Pf(I - P)fPe_k \rangle \\
&= \sum_{k \in \mathbb{Z}} \sum_{n \in \mathbb{Z}} \chi(n + k)\chi(-k - 1) \cdot |f_n|^2 \\
&= \sum_{k \in \mathbb{N}} \sum_{n \in \mathbb{Z}} \chi(n - k) \cdot |f_n|^2 \\
&= \sum_{n \in \mathbb{N}} n \cdot |f_n|^2
\end{aligned}
$$

where we have used Fubinis theorem to interchange the summation since

$$
\begin{aligned}
\sum_{n \in \mathbb{N}} n \cdot |f_n|^2 &\leq \sum_{n \in \mathbb{Z}} n^2 \cdot |f_n|^2 \\
&= \frac{1}{2\pi} \int_0^{2\pi} |f'(\theta)|^2 \, d\theta \\
&< \infty
\end{aligned}
$$

due to the fact that $f \in C^1(S^1)$. We then conclude that $Pf(1 - P)$ and $[P, if]$ is Hilbert-Schmidt. So $if \in u_2(\mathcal{H}, P)$, whence $e^{if} \in \mathcal{U}_2(\mathcal{H}, P)$. Evidently SLS^1 can be realized as a subgroup of $\mathcal{U}_2(\mathcal{H}, P)$ (for an alternative proof see [P-S, p. 82-83]). Notice that we may associate the Lie algebra slS^1, consisting of skew-self-adjoint multiplication operators, if, such that $[P, if]$ is a Hilbert-Schmidt operator, with the special loop group SLS^1. Hence, the *special loop algebra* slS^1 generates SLS^1 through the exponential mapping.

Due to the above inclusion of slS^1 into $u_2(\mathcal{H}, P)$ and SLS^1 into $\mathcal{U}_2(\mathcal{H}, P)$, the spin representation, constructed in sections 2.4 and 2.3, is well-defined and gives a projective representation of SLS^1. We will now show that this spin representation is, in fact, of positive energy. In order to achieve this, we have to construct a Hamilton operator (to apply some meaning to the statement).

Let d_0 denote the generator for rotations in $\mathcal{H} = L_2(S^1)$, i.e. $d_0 = \frac{d}{d\theta}$, in generalized sense, on its maximal domain $\mathcal{D}(d_0) = \{f \in \mathcal{H} : d_0 f \in \mathcal{H}\}$. It follows, by use of Stone's theorem that d_0 is skew-self-adjoint. Even though d_0 is unbounded, its commutator with P still makes sense. Since $[P, d_0]e_k = Pd_0e_k - d_0 Pe_k = 0$, $[P, d_0]$ vanishes on $\mathcal{D}(d_0)$, hence it is well-defined and vanishes on all of $\mathcal{H}$ (as long as we do not consider other unbounded operators, we will not get into trouble). We define a Hamilton operator H on $\mathcal{F}_\wedge(\mathcal{H})$ by $H = -idU_P(d_0)$ with the dense domain $\mathcal{D}(H)$ given below. Note that $[P, d_0] = 0$ and $[\Gamma, d_0]_+ = 0$, where Γ is a given involution on $\mathcal{H}$ commuting with P (see sections 2.3 and 2.2). It follows that

$$\begin{aligned} H &= -idU_P(d_0) = -idU((d_0)_P) \\ &= -idU(d_0(I - 2P)) \end{aligned}$$

where $dU((d_0)_P)$ is defined by the Fock-Cook quantization mapping (see section 2.3), since the skew-self-adjoint unbounded operator d_0 is linear. Notice that $(d_0)_P e_k = d_0(I - 2P)e_k = i|k| \cdot e_k$, for all $k \in \mathbf{Z}$, due to the polarization operator $J = I - 2P$ (which has eigenvalues ± 1 and corresponding eigenspaces $\mathcal{H}_\pm$). Hence, on arbitrary basis product vectors $e_{k_1} \wedge \cdots \wedge e_{k_n}$ in $\wedge^n \mathcal{H}$, $n \in \mathbf{N}$

$$\begin{aligned} &H(e_{k_1} \wedge \cdots \wedge e_{k_n}) \\ &= \sum_{j=1}^n e_{k_1} \wedge \cdots \wedge e_{k_{j-1}} \wedge -i(d_0)_P e_{k_j} \wedge e_{k_{j+1}} \wedge \cdots \wedge e_{k_n} \\ &= \left(\sum_{j=1}^n |k_j|\right) \cdot (e_{k_1} \wedge \cdots \wedge e_{k_n}) \end{aligned}$$

Moreover, $H\Omega = 0$, so the eigenspace corresponding to the eigenvalue 0 is spanned by Ω and e_0, and the eigenspace corresponding to any $m \in \mathbf{N}$ is spanned by basis product vectors $e_{k_1} \wedge \cdots \wedge e_{k_n}$

with $\sum_{j=1}^{n} |k_j| = m$. Observe that these eigenspaces are of finite dimension (each $k_j \in \{-m, -m+1, \ldots, m-1, m\}$ and $n \le m$). The spectrum of H is $\sigma(H) = \mathbf{N} \cup \{0\}$ and evidently $H \ge 0$ (so the representation is indeed of positive energy). Notice that H indeed is unbounded and that the maximal domain

$$\mathcal{D}(H) = \{F \in \mathcal{F}_\wedge(\mathcal{H}) : HF \in \mathcal{F}_\wedge(\mathcal{H})\}$$

for H consists of particles (vectors) of finite energy. $\mathcal{D}(H)$ is evidently dense in $\mathcal{F}_\wedge(\mathcal{H})$, since it includes the algebraic direct sum of the energy eigenspaces and $\mathcal{F}_\wedge(\mathcal{H})$ is the (Hilbert space) direct sum of the energy eigenspaces, i.e. $\mathcal{F}_\wedge(\mathcal{H})$ is the completion of $\mathcal{D}(H)$. Since $f \in C^1(S^1)$, $\|if\|_\infty < \infty$ and the multiplication operator $M(if)$ multiplying with if is bounded. We define $\phi(f)$ to be the self-adjoint closure of the essential self-adjoint generator $-idU_P(if)$, for $if \in slS^1 \subset u_2(\mathcal{H}, P)$. From the commutation relations (2.17) we have

$$
\begin{aligned}
[\phi(f), \phi(g)] &= -[dU_P(if), dU_P(ig)] \\
&= -dU([(if)_P, (ig)_P]) - \omega((if)_P, (ig)_P) \cdot I \\
&= -\omega_P(if, ig) \cdot I
\end{aligned}
$$

on $\mathcal{D}$ since $[(if)_P, (ig)_P] = [if, ig]_P = 0$ and where the Lie algebra cocycle $\omega_P(if, ig) = \omega((if)_P, (ig)_P)$ is given by (2.16) as

$$
\begin{aligned}
\omega_P(if, ig) &= \omega((if)_P, (ig)_P) \\
&= \mathrm{Tr}(Pg(I-P)fP) - \mathrm{Tr}(Pf(I-P)gP)
\end{aligned}
$$

for all $if, ig \in slS^1$. This means that we have constructed a representation of the CCR-algebra (see (3.9) and definition 13 of Chap. 3) in the anti-symmetric Fock Hilbert space $\mathcal{F}_\wedge(\mathcal{H})$, indicating the so-called boson-fermion correspondance. We return to the boson-fermion correspondance in section 5.1.

From section 2.3 it follows that the charge operator $Q = dU(I - 2P)$ commutes with every $dU_P(A)$, $A \in u_2(\mathcal{H}, P)$. This holds especially for $A = if \in slS^1$, i.e. $[i\phi(f), Q] = 0$ on $\mathcal{D}$. Hence, $W(f) = e^{i\phi(f)}$ commutes with the charge operator Q on $\mathcal{D}$, and the (corresponding) representation of the CCR-algebra is highly reducible, since each charge sector $\mathcal{H}_q$ (see section 2.3) is invariant.

Now we complexify the mapping $f \mapsto \phi(f)$ by putting $\phi(f) = \phi(f_r) + i\phi(f_i)$, for $f = f_r + if_i$ in $C^1(S^1)_{\mathbf{C}} = C^1(S^1) + iC^1(S^1)$

such that both f_r and f_i are real and have vanishing trace. Then this extended $\phi(f)$ is no longer self-adjoint, however, $f \mapsto \phi(f)$ is a "*-quantization mapping", i.e. $\phi(f)^* = \phi(\bar{f})$, on $\mathcal{D}$, since $\langle \phi^*(f)F_n, G_m \rangle = \langle F_n, (\phi(f_r) + i\phi(f_i))G_m \rangle = \langle (\phi(f_r) - i\phi(f_i)F_n, G_m \rangle = \langle \phi(\bar{f})F_n, G_m \rangle$, for any $F_n \in \wedge^n \mathcal{H}$ and $G_m \in \wedge^m \mathcal{H}$. Moreover, each $f \in C^1(S^1)_{\mathbb{C}}$ can be uniquely decomposed as $f = f_+ \oplus f_-$ where $f_+ = \sum_{k \in \mathbf{N}} \langle e_k, f \rangle e_k \in \mathcal{H}_+$ and $f_- = \sum_{k \in -\mathbf{N}} \langle e_k, f \rangle e_k \in \mathcal{H}_-$. Recall that $f_0 = 0$ due to the vanishing trace, notice also that $\langle f_+, f_- \rangle = 0$. Then $\phi(f) = \phi(f_+) + \phi(f_-)$. The observation

$$
\begin{aligned}
\overline{(f_-)} &= \overline{\sum_{k \in -\mathbf{N}} \langle e_k, f \rangle e_k} = \sum_{-k \in \mathbf{N}} \langle e_{-k}, \bar{f} \rangle e_{-k} \\
&= \sum_{k \in \mathbf{N}} \langle e_k, \bar{f} \rangle e_k = (\bar{f})_+
\end{aligned}
$$

implies that $\overline{(f_-)} = (\bar{f})_+ = f_+$ in case f is real (notice that f_+ and f_- may be non-real even though f itself is real). Hence, $\phi(f_-)^* = \phi(f_+)$ for $if \in slS^1$, and

$$
\begin{aligned}
\|\phi(f_-)\Omega\|^2 &= \langle \Omega, \phi(f_+)\phi(f_-)\Omega \rangle = \mathrm{Tr}(Pf_+(I - P)f_- P) \\
&= \sum_{m>0} \sum_{n<0} \sum_{k \in \mathbf{Z}} f_m f_n \chi(-k - 1)\chi(k + n)\delta_{m+n} \\
&= \sum_{n<0} \sum_{k<0} |f_n|^2 \cdot \chi(k + n) = 0
\end{aligned}
$$

then $\phi(f_-)\Omega = 0$, for $if \in slS^1$.

We will now compute the cocycle $\omega_P(if, ig)$ explicitly. With the earlier notation we have $Pe_k = \chi(-k - 1)e_k$, so

$$
gPe_k = \sum_{n \in \mathbf{Z}} g_n \chi(-k - 1)e_{n+k}
$$

and

$$
f(I - P)gPe_k = \sum_{m \in \mathbf{Z}} \sum_{n \in \mathbf{Z}} f_m g_n \chi(-k - 1)\chi(k + n)e_{k+n+m}
$$

Hence

$$
\begin{aligned}
&\mathrm{Tr}(Pf(I - P)gP) \\
&= \sum_{k \in \mathbf{Z}} \langle Pe_k, f(I - P)gPe_k \rangle
\end{aligned}
$$

$$= \sum_{k\in\mathbf{Z}}\sum_{m\in\mathbf{Z}}\sum_{n\in\mathbf{Z}} f_m g_n \chi(-k-1)\chi(k+n)\cdot\delta_{m+n}$$

$$= \sum_{k\in-\mathbf{N}}\sum_{m\in\mathbf{Z}} f_m g_{-m}\chi(k-m)$$

$$= \sum_{k\in-\mathbf{N}}\sum_{n\in\mathbf{Z}} \overline{f_n} g_n \chi(k+n)$$

$$= \sum_{n\in\mathbf{N}} n\overline{f_n} g_n$$

where we have used Fubini's theorem to interchange the summations, since

$$n\cdot|\overline{f_n}|\cdot|g_n| \le \frac{1}{2}\left(n^2\cdot|\overline{f_n}|^2 + |g_n|^2\right)$$

for $n\in\mathbf{N}$, and

$$\sum_{n\in\mathbf{N}} n\cdot|\overline{f_n}|\cdot|g_n| \le \frac{1}{2}\sum_{n\in\mathbf{N}} n^2\cdot|\overline{f_n}|^2 + \frac{1}{2}\sum_{n\in\mathbf{N}}|g_n|^2$$

$$= \frac{1}{4\pi}\int_0^{2\pi}|f'(\theta)|^2\,d\theta + \frac{1}{4\pi}\int_0^{2\pi}|g(\theta)|^2\,d\theta$$

which is finite, because $f,g\in C^1(S^1)$. Hence

$$\omega_P(if,ig) = \sum_{n\in\mathbf{N}} n\cdot\left(f_n\cdot\overline{g_n} - \overline{f_n}\cdot g_n\right)$$

for $if,ig\in slS^1$, where $f_n = \langle e_n,f\rangle = \frac{1}{2\pi}\int_0^{2\pi} f(\theta)e^{-in\theta}\,d\theta$. Furthermore

$$\omega_P(if,ig) = \frac{1}{2\pi i}\int_0^{2\pi} f'(\theta)g(\theta)d\theta$$

for $if,ig\in slS^1$, since

$$\frac{1}{2\pi i}\int_0^{2\pi} f'(\theta)g(\theta)d\theta = \sum_{n\in\mathbf{Z}} n\cdot f_n\cdot\overline{g_n}$$

$$= \sum_{n\in\mathbf{N}} n\cdot\left(f_n\cdot\overline{g_n} - \overline{f_n}\cdot g_n\right)$$

due to $f_0 = g_0 = 0$, $f'(\theta) = \sum_{n\in\mathbf{Z}} in f_n e_n(\theta)$ and

$$n f_n\overline{g_n} = n\overline{f_{-n}}g_{-n} = -(-n)\overline{f_{-n}}g_{-n} = -k\overline{f_k}g_k$$

for $k = -n$ and n negative. Notice that $\omega_P(\cdot, \cdot)$ evidently defines a non-degenerated symplectic form on $slS^1 \times slS^1$, since

$$
\begin{aligned}
\omega_P(if, ig) + \omega_P(ig, if) &= \frac{1}{2\pi i} \int_0^{2\pi} (f(\theta)g(\theta))' d\theta \\
&= \frac{1}{2\pi i} [f(\theta)g(\theta)]_0^{2\pi} \\
&= 0
\end{aligned}
$$

giving the anti-symmetry of the form, and $\int_0^{2\pi} f'(\theta)g(\theta)d\theta = 0$, for all $ig \in slS^1$, implies that $f'(\theta) = 0$ so $f(\theta) = f_0 = 0$, giving the non-degeneracy of the form.

The so-called two-point function becomes

$$
\begin{aligned}
\langle \Omega, \phi(f)\phi(g)\Omega \rangle &= -\langle \Omega, dU_P(if)dU_P(ig)\Omega \rangle \\
&= -\mathrm{Tr}(Pif(I - P)igP) \\
&= \sum_{n \in \mathbf{N}} n \cdot \overline{f_n}g_n
\end{aligned}
$$

by use of formula (2.15) derived in the end of section 2.3 and the above trace formula.

The functions in $i \cdot slS^1$, considered as real functions on S^1 whose integrals vanish, are completely determined by their positive Fourier coefficients. In fact, the Fourier transformation $F : f \to (f_k)_{k \in \mathbf{N}}$ is an isomorphism. On the space of all these Fourier transforms $(f_k)_{k \in \mathbf{N}}$, $f \in i \cdot slS^1$

$$
\langle (f_k)_{k \in \mathbf{N}}, (g_n)_{n \in \mathbf{N}} \rangle = \sum_{k \in \mathbf{N}} k \cdot \overline{f_k} \cdot g_k
$$

is evidently a complex inner product, whereby $F(i \cdot slS^1)$ becomes a pre-Hilbert space. Hence, we may turn $i \cdot slS^1$ into a real pre-Hilbert space, with inner product given by the real part of

$$
\langle f, g \rangle_{1/2} = \sum_{k \in \mathbf{N}} k \overline{f_k} \cdot g_k
$$

for $f, g \in i \cdot slS^1$. We denote the corresponding real Hilbert space by $\mathcal{H}_r^{1/2}$. Furthermore, we let $\mathcal{H}^{1/2}$ denote the corresponding complex Hilbert space defined by introducing the ordinary complex structure on the Fourier components. We remark that the index and super-script 1/2 used above, serve to indicate the analogy with the Sobolev space of a "half time" differentiable functions (in generalized sense). Observe that we then get the identity $\omega_P(if, ig) = -2i \cdot Im \langle f, g \rangle_{1/2}$.

The spin representation of the special loop group SLS^1, is given by $U_P(e^{if}) = e^{i \cdot \phi(f)}$. It leaves each charge sector $\mathcal{H}_q$ invariant and it fulfils the Weyl form of the canonical commutation relations

$$U_P \left(e^{if} \right) U_P \left(e^{ig} \right) = e^{\frac{1}{2}\omega(if,ig)} U_P \left(e^{i(f+g)} \right)$$

for all $f, g \in \mathcal{H}_r^{1/2}$, as we will show below. Later on we show that these representations on different charge sectors $\mathcal{H}_q$, are unitarily equivalent.

Recall that $\mathcal{D}$ is a dense set of analytical vectors for $\phi(f) = -i \cdot dU_P(if)$ and since it is closed, by definition, it is self-adjoint. Let $F, G \in \mathcal{D}$ be arbitrarily chosen, then

$$\langle \phi(g)G, e^{-i\phi(f)}F \rangle = \sum_{n=0}^{\infty} \left\langle \phi(g)G, \frac{(-i)^n}{n!} \phi(f)^n F \right\rangle$$

$$= \sum_{n=0}^{\infty} \left\langle G, \frac{(-i)^n}{n!} (\phi(f)^n \phi(g) + n \cdot \phi(f)^{n-1} \omega_P(if, ig)) F \right\rangle$$

where we have used the self-adjointness of $\phi(g)$ combined with the fact that $[\phi(g), \phi(f)^n] = n \cdot \omega_P(if, ig) \cdot \phi(f)^{n-1}$ due to the commutation relations derived above. Consequently

$$\langle \phi(g)G, e^{-i\phi(f)}F \rangle$$

$$= \sum_{n=0}^{\infty} \left\langle \frac{i^n}{n!} \phi(f)^n G, \phi(g)F \right\rangle + \langle G, -ie^{-i\phi(f)} \omega_P(if, ig)F \rangle$$

$$= \langle G, e^{-i\phi(f)}(\phi(g) - i \cdot \omega_P(if, ig))F \rangle$$

Since $e^{-i\phi(f)}$ is bounded, in fact it is unitary, it follows that $\phi(g)^* = \phi(g)$ is well-defined on $e^{-i\phi(f)}F$ and

$$\phi(g)e^{-i\phi(f)}F = e^{-i\phi(f)}(\phi(g) - i \cdot \omega_P(if, ig) \cdot I)F$$

for any $F \in \mathcal{D}$. Thus

$$\left\| \phi(g)e^{-i\phi(f)}F \right\| \le \|\phi(g)F\| + |\omega_P(if, ig)| \cdot \|F\|$$

where the unitarity of $e^{-i\phi(f)}$ has been used. Since $\mathcal{D}$ is a core for $\phi(g)$, i.e. $\overline{\phi(g)\,|_{\mathcal{D}}} = \phi(g)$, any $F \in \mathcal{D}(\phi(g))$ can be approximated by a sequence $\{F_k\}_{k\in\mathbb{N}} \subset \mathcal{D}$, such that $F_k \to F$ and $\phi(g)F_k \to \phi(g)F$. Then

$$\|\phi(g)e^{-i\phi(f)}(F_n - F_m)\|$$
$$= \|\phi(g)(F_n - F_m)\| + |\omega_P(if, ig)| \cdot \|F_n - F_m\|$$
$$\to 0 \quad \text{as } m, n \to \infty$$

showing that $\{\phi(g)e^{-i\phi(f)}F_k\}_{k\in\mathbb{N}}$ form a Cauchy sequence and hence converge. Therefore $e^{-i\phi(f)}F \in \mathcal{D}(\phi(g))$ and

$$\phi(g)e^{-i\phi(f)}F_k \to \phi(g)e^{-i\phi(f)}F$$

Thus, $\mathcal{D}(\phi(g))$ is invariant under each $e^{-i\phi(f)}$ and the formula

$$\phi(g)e^{-i\phi(f)} = e^{-i\phi(f)}(\phi(g) - i \cdot \omega_P(if, ig) \cdot I)$$

extend to all of $\mathcal{D}(\phi(g))$. Then, for any $F \in \mathcal{D}(\phi(g))$

$$\frac{d}{ds}e^{is\phi(f)}e^{is\phi(g)}e^{-is\phi(f+g)}F$$
$$= e^{is\phi(f)}e^{is\phi(g)}i \cdot (\phi(f) + \phi(g) - \phi(f+g))e^{-is\phi(f+g)}F$$
$$\quad + e^{is\phi(f)}i \cdot (i \cdot \omega_P(isg, if))e^{is\phi(g)}e^{-is\phi(f+g)}F$$
$$= s \cdot \omega_P(if, ig)e^{is\phi(f)}e^{is\phi(g)}e^{-is\phi(f+g)}F$$

since $\omega_P(ig, if) = -\omega_P(if, ig)$. Notice that all of the products above are well-defined. Integration then gives the following identity between bounded operators

$$e^{i\phi(f)}e^{i\phi(g)}e^{-i\phi(f+g)} = I + \int_0^1 ds\,\omega_P(if, ig) \cdot s \cdot e^{is\phi(f)}e^{is\phi(g)}e^{-is\phi(f+g)}$$

This equation can be solved by iteration as follows. Put $V_0(f, g) = I$ and recursively $V_{n+1}(f, g) = \int_0^1 ds\,\omega_P(if, ig) \cdot s \cdot V_n(sf, sg)$, for $n \in \mathbb{N} \cup \{0\}$. Then

$$V_1(f, g) = \int_0^1 ds\,\omega_P(if, ig) \cdot s \cdot I = \frac{\omega_P(if, ig)}{2} \cdot I$$

Assume that $V_n(f, g) = \left(\frac{\omega_P(if, ig)}{2}\right)^n \cdot \frac{1}{n!} \cdot I$, then $V_n(sf, sg) = s^{2n} \cdot V_n(f, g)$ and

$$V_{n+1}(f, g) = \int_0^1 ds\,\omega_P(if, ig) \cdot s \cdot s^{2n} \cdot V_n(f, g)$$
$$= \left(\frac{\omega_P(if, ig)}{2}\right)^{n+1} \cdot \frac{1}{(n+1)!} \cdot I$$

So induction gives that the formula above holds for all $n \in \mathbb{N} \cup \{0\}$.
Define

$$V(f,g) = \lim_{N \to \infty} \sum_{n=0}^{N} V_n(f,g) = e^{\frac{1}{2}\omega_P(if,ig)} \cdot I$$

Hence, $V(f,g)$ is the unique solution to the integral equation

$$
\begin{aligned}
V(f,g) &= \sum_{n=0}^{\infty} V_n(f,g) \\
&= I + \int_0^1 ds \cdot \omega_P(if,ig) \cdot s \cdot \sum_{n=1}^{\infty} V_{n-1}(sf,sg) \\
&= I + \int_0^1 ds \cdot \omega_P(if,ig) \cdot s \cdot V(sf,sg)
\end{aligned}
$$

whence

$$e^{i\phi(f)} e^{i\phi(g)} e^{-i\phi(f+g)} = V(f,g) = e^{\frac{1}{2}\omega_P(if,ig)} \cdot I$$

and

$$
\begin{aligned}
e^{i\phi(f)} e^{i\phi(g)} &= e^{\frac{1}{2}\omega_P(if,ig)} \cdot e^{i\phi(f+g)} \\
&= e^{\omega_P(if,ig)} \cdot e^{i\phi(g)} e^{i\phi(f)}
\end{aligned}
$$

for any $f,g \in i \cdot slS^1$.

We end this discussion of the special loop group by calculating the vacuum functional. The formula

$$\phi(g)\phi(f)^m = \phi(f)^m \phi(g) + m \cdot \omega_P(if,ig)\phi(f)^{m-1}$$

on $\mathcal{D}$, almost holds with $g = f_-$, even though f_- is complex, since it holds for its real and imaginary parts f_-^r and f_-^i, both in $C^1(S^1)_r$, where $f_- = f_-^r + i \cdot f_-^i$. Thus, direct calculations give

$$
\begin{aligned}
\phi(f_-)\phi(f)^m F &= (\phi(f_-^r) + i \cdot \phi(f_-^i))\phi(f)^m F \\
&= \phi(f)^m (\phi(f_-^r) + i \cdot \phi(f_-^i))F \\
&\quad + m \cdot (\omega_P(if,if_-^r) + i \cdot \omega_P(if,if_-^i))\phi(f)^{m-1} F \\
&= \phi(f)^m \phi(f_-)F + m \cdot \|f\|_{1/2}^2 \phi(f)^{m-1} F
\end{aligned}
$$

where $\|f\|_{1/2}^2 = \langle f, f \rangle_{1/2}$. For $F = \Omega$ we obtain

$$\phi(f_-)\phi(f)^m \Omega = m \cdot \|f\|_{1/2}^2 \cdot \phi(f)^{m-1}\Omega$$

since $\phi(f_-)\Omega = 0$. Because Ω is an analytic vector for $\phi(f)$, we have

$$
\begin{aligned}
\langle \Omega, U_P\left(e^{if}\right)\Omega \rangle &= \langle \Omega, e^{i\phi(f)}\Omega \rangle \\
&= \sum_{n=0}^{\infty} \frac{i^n}{n!} \langle \Omega, \phi(f)^n\Omega \rangle
\end{aligned}
$$

We obtain immediately

$$
\begin{aligned}
\langle \Omega, \phi(f)^n\Omega \rangle &= \langle \Omega, (\phi(f_+) + \phi(f_-))\phi(f)^{n-1}\Omega \rangle \\
&= \langle \phi(f_-)\Omega, \phi(f)^{n-1}\Omega \rangle + \langle \Omega, \phi(f_-)\phi(f)^{n-1}\Omega \rangle \\
&= 0 + \left\langle \Omega, (n-1)\cdot \|f\|_{1/2}^2 \cdot \phi(f)^{n-2}\Omega \right\rangle \\
&= (n-1)\cdot \|f\|_{1/2}^2 \cdot \langle \Omega, \phi(f)^{n-2}\Omega \rangle
\end{aligned}
$$

Hence, by induction we have

$$
\langle \Omega, \phi(f)^n\Omega \rangle = 0
$$

for $n \in \mathbf{N}$ odd, since $\langle \Omega, \phi(f)\Omega \rangle = 0$, and .

$$
\begin{aligned}
\langle \Omega, \phi(f)^n\Omega \rangle &= (n-1)(n-3)\cdot \ldots \cdot 1 \cdot \|f\|_{1/2}^n \\
&= \frac{n!}{\left(\frac{n}{2}\right)!\cdot 2^{n/2}} \cdot \|f\|_{1/2}^n
\end{aligned}
$$

for $n \in \mathbf{N}$ even, since $\langle \Omega, \Omega \rangle = 1$. Thus

$$
\begin{aligned}
\langle \Omega, U_P\left(e^{if}\right)\Omega \rangle &= \sum_{k=0}^{\infty} \frac{i^{2k}}{(2k)!} \frac{(2k)!}{k!\cdot 2^k} \cdot \|f\|_{1/2}^{2k} \\
&= \sum_{k=0}^{\infty} \frac{\left(-\frac{1}{2}\right)^k}{k!} \cdot \left(\|f\|_{1/2}^2\right)^k \\
&= e^{-\frac{1}{2}\cdot\|f\|_{1/2}^2}
\end{aligned}
$$

giving the vacuum functional.

We now turn to the charge subgroup C, of the loop group LS^1, generated by elements of the form e^{iF_0}, where $F_0(\theta) = n_F \cdot \theta + f_0$, with $n_F \in \mathbf{Z}$ and $f_0 \in \mathbf{R}$.

The charge group C, as a subgroup of LS^1, has infintely many disconnected components and these components are labelled by an integer n_F, the winding number for e^{iF}.

The shift operator s on $\mathcal{H}$ given by $s(e_k) = e_{k+1}$, i.e. $(sh)(\theta) = e^{i\theta} \cdot h(\theta)$, for any $h \in \mathcal{H}$, is evidently unitary. Hence, $s \in \mathcal{U}(\mathcal{H})$, in fact, $s \in \mathcal{U}_2(\mathcal{H}, P)$ since $[P, s]$ is a Hilbert-Schmidt operator, due to $[P, s]e_k = Pe_{k+1} - sPe_k = -\delta_{k+1} \cdot e_0$, so $[P, s]$ is the rank 1 operator $f \rightarrow -\langle e_{-1}, f \rangle e_0$. However, the shift operator s is not generated by any element in the Lie algebra $u_2(\mathcal{H}, P)$, since its generator $i\theta$, commuting with P, does not fulfil the Hilbert-Schmidt condition. We will show that $[P, \theta]$ has an infinite Hilbert-Schmidt norm. Let $b(\theta)$ denote the function $\theta \mapsto \theta$, on $[0, 2\pi]$. The Fourier series is given by $b(\theta) = \sum_{k \in \mathbf{Z}} b_k e_k(\theta)$, where $b_0 = \pi$ and $b_k = \frac{i}{k}$,, for $k \neq 0$. Hence

$$
\begin{aligned}
[P, \theta]e_n &= P\sum_{k \in \mathbf{Z}} b_k e_{k+n} - \sum_{k \in \mathbf{Z}} b_k e_{k+n} \chi(-n-1) \\
&= \sum_{k \in \mathbf{Z}} b_k e_{k+n} \cdot (\chi(-k-n-1) - \chi(-n-1)) \\
&= \begin{cases} \sum_{k < -n} b_k e_{k+n} & , \text{ for } n \geq 0 \\ -\sum_{k \geq -n} b_k e_{k+n} & , \text{ for } n < 0 \end{cases}
\end{aligned}
$$

where χ denotes the indicator function for $\mathbf{N} \cup \{0\}$ defined above. So

$$
\begin{aligned}
\| [P, \theta] \|_{HS}^2 &= \sum_{n \in \mathbf{Z}} \| [P, \theta]e_n \|^2 \\
&= \sum_{n \in \mathbf{N} \cup \{0\}} \sum_{k < -n} |b_k|^2 + \sum_{n \in -\mathbf{N}} \sum_{k \geq -n} |b_k|^2 \\
&= \sum_{n \in \mathbf{N} \cup \{0\}} \sum_{k < -n} \frac{1}{k^2} + \sum_{n \in -\mathbf{N}} \sum_{k \geq -n} \frac{1}{k^2} \\
&= 2 \cdot \sum_{n \in \mathbf{N}} \sum_{k \geq n} \frac{1}{k^2} \\
&> 2 \cdot \sum_{n \in \mathbf{N}} \int_n^\infty \frac{1}{x^2} dx = 2 \cdot \sum_{n \in \mathbf{N}} \frac{1}{n}
\end{aligned}
$$

which is a well-known divergent series. Therefore $[P, \theta]$ is not a Hilbert-Schmidt operator, as stated. This means that we cannot use the method developed in Chap. 2 to find a unitary operator $U_P(s)$, such that

$$
a_P(sh) = U_P(s)a_P(h)U_P(s)^{-1}
$$

for all $h \in \mathcal{H}$. In this case one can nevertheless explicitly construct such a unitary operator $U_P(s)$. Notice that $a_P(e_k) = a((1 - P)e_k) +$

$a^*(\Gamma P e_k) = \chi(k) \cdot a(e_k) + \chi(-k-1) \cdot a^*(e_k)$, so $a_P(e_k) = a(e_k)$, for $k \geq 0$, and $a_P(e_k) = a^*(e_k)$, for $k < 0$, and then $a_P(s e_k) = a_P(e_{k+1}) = a(e_{k+1})$, for $k \geq -1$, and $a_P(s e_k) = a_P(e_{k+1}) = a^*(e_{k+1})$, for $k < -1$. Hence, the condition for $U_P(s)$ is equivalent to

$$U_P(s) a(e_k) U_P(s)^{-1} = a(e_{k+1})$$

for $k \neq -1$ and

$$U_P(s) a(e_{-1}) U_P(s)^{-1} = a^*(e_0)$$

Define an operator S on $\mathcal{F}_\wedge(\mathcal{H})$ by its action on product basis vectors $S\Omega = e_0$, $S e_{-1} = \Omega$, $S(e_{k_1} \wedge \ldots \wedge e_{k_n}) = e_{k_1+1} \wedge \ldots \wedge e_{k_n+1} \wedge e_0$, provided each $k_j \neq -1$, for $j = 1, \ldots, n$, and $S(e_{k_1} \wedge \ldots \wedge e_{k_n} \wedge e_{-1}) = e_{k_1+1} \wedge \ldots \wedge e_{k_n+1}$, where each $k_j \neq -1$, for $j = 1, \ldots, n$. Extension by linearity and continuity gives a bounded operator S on $\mathcal{F}_\wedge(\mathcal{H})$. From above it follows that S is a surjective isometry, hence it is unitary. Let Q be the charge operator on $\mathcal{F}_\wedge(\mathcal{H})$, with domain $\mathcal{D}$, defined in section 2.3. From the action of S on product basis vectors, which belongs to $\mathcal{D}$, we immediately see that S maps the q'th charge sector $\mathcal{H}_q$ into the $(q+1)$'th charge sector $\mathcal{H}_{q+1}$. Then $(QS - SQ)(e_{k_1} \wedge \ldots \wedge e_{k_n}) = (q+1-q)S(e_{k_1} \wedge \ldots \wedge e_{k_n})$, where $e_{k_1} \wedge \ldots \wedge e_{k_n}$ denotes an arbitrary product basis vector in $\mathcal{H}_q$ (it also holds on $\Omega \in \mathcal{H}_0$). We then get the following commutation relation $[Q, S] = S$ on $\mathcal{D}$, or equivalently

$$QS = S(Q + I)$$

By direct calculation it follows that S satisfies our condition to $U_P(s)$ and we may put $U_P(s) = S$. Hence, we have constructed the unitary operator $U_P(s)$ such that $a_P(sh) = U_P(s) a_P(h) U_P(s)^{-1}$, for all $h \in \mathcal{H}$. Notice that s^n acts as the multiplication operator $e^{in\theta}$. We therefore define $U_P(s^n) = S^n$, hence $a_P(s^n h) = S^n a_P(h)(S^{-1})^n = U_P(s^n) a_P(h) U_P(s^n)^{-1}$, for all $h \in \mathcal{H}$ and $n \in \mathbf{Z}$. Now we have handled the first term $n_F \cdot \theta$ in F_0. We thus turn to the second term f_0 in $F_0 = n_F \cdot \theta + f_0$.

Consider e^{if_0}, with $f_0 \in \mathbf{R}$, which trivially belongs to the restricted unitary group $\mathcal{U}_2(\mathcal{H}, P)$. Then the unitary operator $U_P(e^{if_0})$ on $\mathcal{F}_\wedge(\mathcal{H})$ is explicitly given by $U_P(e^{if_0}) = e^{dU_P(if_0)} = e^{f_0 \cdot dU_P(iI)} = e^{if_0 Q}$, where Q is the change operator, discussed in section 2.3.

Combining the above discussion of the two terms in F_0, we are lead to define $U_P(e^{iF_0})$ by

$$U_P\left(e^{iF_0}\right) = e^{\frac{1}{2}if_0 Q} S^{n_F} e^{\frac{1}{2}if_0 Q}$$

Then $U_P\left(e^{iF_0}\right)$ is unitary and $U_P(e^{iF_0})a_P(h)U_P(e^{iF_0})^{-1} = a_P(e^{iF_0}h)$, as intend.

We will now calculate an explicit formula for the associated co-cycle. For arbitrary $n \in \mathbf{N} \cup \{0\}$, is $QS^n = S^n(Q + n \cdot I)$ and $Q^k S^n = S^n(Q + n \cdot I)^k$ on $\mathcal{D}$. Then $e^{\lambda Q}S^n = S^n e^{\lambda(Q+n\cdot I)}$ on $\mathcal{D}$, for any $n \in \mathbf{N} \cup \{0\}$, hence any $n \in \mathbf{Z}$, and any $\lambda \in \mathbf{C}$, since $\mathcal{D}$ is a set of analytic vectors for Q (see section 2.3), which is invariant under S. Hence

$$e^{\frac{1}{2}if_0 Q}S^{n_G} = S^{n_G}e^{\frac{1}{2}if_0(Q+n_G)} = S^{n_G}e^{\frac{1}{2}if_0 Q}e^{\frac{1}{2}if_0 n_G}$$

and

$$S^{n_F}e^{\frac{1}{2}ig_0 Q} = e^{\frac{1}{2}ig_0 Q}S^{n_F}e^{-\frac{1}{2}ig_0 n_F}$$

which implies

$$\begin{aligned} U_P&\left(e^{iF_0}\right)U_P\left(e^{iG_0}\right)\\ &= e^{\frac{1}{2}if_0 Q}S^{n_F}e^{\frac{1}{2}if_0 Q}e^{\frac{1}{2}ig_0 Q}S^{n_G}e^{\frac{1}{2}ig_0 Q}\\ &= e^{\frac{1}{2}if_0 Q}e^{\frac{1}{2}ig_0 Q}e^{-\frac{1}{2}ig_0 n_F}S^{n_F}S^{n_G}e^{\frac{1}{2}if_0 Q}e^{\frac{1}{2}ig_0 Q}e^{\frac{1}{2}if_0 n_G}\\ &= e^{\frac{1}{2}i(f_0 n_G - g_0 n_F)}e^{\frac{1}{2}i(f_0+g_0)Q}S^{(n_F+n_G)}e^{\frac{1}{2}i(f_0+g_0)Q}\\ &= c\left(e^{iF_0}, e^{iG_0}\right)\cdot U_P\left(e^{i(F_0+G_0)}\right) \end{aligned}$$

whereby we see that the cocycle is given by

$$c\left(e^{iF_0}, e^{iG_0}\right) = e^{\frac{1}{2}i(f_0 n_G - g_0 n_F)}$$

Thus, we have constructed a projective unitary representation of the abelian charge group C.

We will now show that $U_P(e^{if})$ and $U_P(e^{iF_0})$ commute for all $if \in slS^1$ and $e^{iF_0} \in C$. This is a consequence of the fact that the unitary operators $V_1 = Se^{i\phi(f)}$ and $V_2 = e^{i\phi(f)}S$ implement the same automorphism. We have

$$\begin{aligned} \alpha_1(a_P(g)) &= V_1 a_P(g)V_1^* = U_P(s)a_P\left(e^{if}g\right)U_P(s)\\ &= a_P\left(e^{i\theta}e^{if}g\right) = a_P\left(e^{i(f+\theta)}g\right) \end{aligned}$$

and

$$\begin{aligned} \alpha_2(a_P(g)) &= V_2 a_P(g)V_2^* = U_P\left(e^{if}\right)a_P(sg)U_P\left(e^{if}\right)^{-1}\\ &= a_P\left(e^{if}e^{i\theta}g\right) = a_P\left(e^{i(f+\theta)}g\right) \end{aligned}$$

so $\alpha_1(a_P(g)) = \alpha_2(a_P(g))$. Since the representation of the CAR-algebra, labelled by P, is irreducible, it follows that $V_1 = c_0(V_1, V_2)V_2$, where $c_0(V_1, V_2) \in \mathbf{C}$ and has absolute value 1, due to the unitarity of V_1 and V_2. That is

$$Se^{i\phi(f)} = c_0(f) \cdot e^{i\phi(f)}S$$

with $|c_0(f)| = 1$, for all $f \in \mathcal{H}$. We write $c_0(f)$ as $e^{ic(f)}$, with $c(f)$ real. Then the unitarity of S gives

$$e^{i(\phi(f)+c(f))} = Se^{i\phi(f)}S^{-1} = e^{iS\phi(f)S^{-1}}$$

on $\mathcal{D}$, since S leaves $\mathcal{D}$ invariant and $\mathcal{D}$ consists of analytic vectors for $\phi(f)$. Putting $f = t \cdot g$ and taking the derivative at $t = 0$ we obtain

$$\phi(g) + c(g) \cdot I = S\phi(g)S^{-1}$$

where we have used that $t \to c(tg)$ is linear. By taking the vacuum expectation value of the equation

$$c(g) \cdot I = \phi(g) - S^{-1}\phi(g)S$$

we get

$$c(g) = \langle \Omega, \phi(g)\Omega \rangle - \langle S\Omega, \phi(g)S\Omega \rangle = 0 - \langle e_0, \phi(g)e_0 \rangle$$

by (2.10). Then

$$
\begin{aligned}
c(g) &= -\langle e_0, idU_P(ig)e_0 \rangle \\
&= i \cdot \langle e_0, ((I-P)ig(I-P) + \Gamma Pig P\Gamma)e_0 \rangle
\end{aligned}
$$

since it is only the linear part of $(ig)_P$ which contributes, due to $dU(((ig)_P)_2)e_0 \in \wedge^3\mathcal{H} \perp \mathcal{H}$, and $dU(((ig)_P)_1)$ is given by the Fock-Cook generator. Hence

$$
\begin{aligned}
c(g) &= -\langle (I-P)e_0, g(I-P)e_0 \rangle + \langle ig\Gamma Pe_0, \Gamma Pe_0 \rangle \\
&= -\langle e_0, ge_0 \rangle + 0 \\
&= 0
\end{aligned}
$$

for any $g \in \mathcal{H}$, since $g_0 = \langle e_0, ge_0 \rangle = 0$. Whereby

$$\phi(f)S = S\phi(f)$$

on $\mathcal{D}$, and

$$e^{i\phi(f)} S = S e^{i\phi(f)}$$

for any $f \in \mathcal{H}$. Thus

$$
\begin{aligned}
U_P\left(e^{if}\right) U_P\left(e^{iF_0}\right) &= e^{i\phi(f)} e^{\frac{1}{2} i f_0 Q} S^{n_F} e^{\frac{1}{2} i f_0 Q} \\
&= e^{\frac{1}{2} i f_0 Q} S^{n_F} e^{\frac{1}{2} i f_0 Q} e^{i\phi(f)} \\
&= U_P\left(e^{iF_0}\right) U_P\left(e^{if}\right)
\end{aligned}
$$

due to the fact that $[Q, e^{i\phi(f)}] = 0$, on $\mathcal{D}$, shown in section 2.3. Hence we put $U_P(e^{iF}) = U_P(e^{iF_0})U_P(e^{if})$, which is the declared projective representation of the loop group LS^1. Notice that all representations of SLS^1, in the different charge sectors, are unitarily equivalent, $S^{-1}\phi(f)|_{\mathcal{H}_{q+1}} S = \phi(f)|_{\mathcal{H}_q}$, since $S : \mathcal{H}_q \to \mathcal{H}_{q+1}$ is unitary.

We have constructed a represenation of the canonical commution relation in each charge sector $\mathcal{H}_q$, by use of the mapping $f \to \phi(f)$, on the anti-symmetric Fock Hilbert space. All these representations are shown to be unitarily equivalent and are, moreover, unitarily equivalent to the Fock representations $f \to \pi(f)$ in the symmetric Fock Hilbert space (modelled over $\mathcal{H}^{1/2}$), discussed in section 3.1. However, we delay the proof of the last claimed equivalence until section 5.1.

5.2 The Diffeomorphism Group $Diff^+(S^1)$ as a Unitary Group

In this section we will study the group consisting of orientation preserving diffeomorphisms of the unit circle S^1. We will do this by realizing $Diff^+(S^1)$ as a subgroup of the restricted unitary group, using the spin representation on a Lie algebra level. It turns out that we get representations of the Virasoro algebra in terms of the spin representation, whereby the connection with Chap. 1 and Chap. 3 appears. However, this is not the only possible realization of $Diff^+(S^1)$.

In section 5.3 we will realize $Diff^+(S^1)$ as a subgroup of the restricted symplectic group, using the metaplectic representation on a Lie algebra level. Thereby we construct positive energy projective representations of the Virasoro algebra. The Lie algebra consists of unbounded operators, which complicates the subject. We consider therefore only relative simple problems. From this the relation between the metaplectic representation and the Virasoro algebra will also become clear.

There is an extensive litterature on the Virasoro algebra; however, the following is mostly based on section 2.3, 3.3 and 4.2 of this book, but also [P-S], [Ne], [Lu 2], [K-R] and [Mi] are of interest for this subject.

Let $Diff^+(S^1)$ denote the orientation preserving diffeomorphism of the unit circle S^1, i.e. an element $\psi \in Diff^+(S^1)$ is of the form $\psi\left(e^{i\theta}\right) = e^{i\cdot\phi(\theta)}$, where $e^{i\theta} \in S^1$, ϕ is a smooth real function such that $\phi(\theta + 2\pi) = \phi(\theta) + 2\pi$ and $\phi'(\theta) > 0$.

In this section we consider the case where $Diff^+(S^1)$ is realized as a subgroup of the restricted unitary group.

The diffeomorphism group $Diff^+(S^1)$ can act on $\mathcal{H} = L_2(S^1)$ in more than one way ([P-S, p.91] and [Ne, p. 411]). The action becomes unitary if we choose it to be

$$(u_\phi f)(\theta) = f(\phi(\theta)) \cdot |\phi'(\theta)|^{\frac{1}{2}}$$

for any $f \in \mathcal{H}$. Notice that $|\phi'(\theta)| = \phi'(\theta)$, so we may omit the absolute value symbol if we want to. Evidently

$$
\begin{aligned}
\langle u_\phi f, u_\phi g \rangle &= \frac{1}{2\pi} \int_0^{2\pi} \overline{f(\phi(\theta))} g(\phi(\theta)) \cdot |\phi'(\theta)| \, d\theta \\
&= \frac{1}{2\pi} \int_{\phi(0)}^{\phi(0)+2\pi} \overline{f(\tilde{\theta})} g(\tilde{\theta}) d\tilde{\theta} \\
&= \langle f, g \rangle
\end{aligned}
$$

for all $f, g \in \mathcal{H}$, since $\phi(\theta + 2\pi) = \phi(\theta) + 2\pi$ and $\phi'(\theta) > 0$. So u_ϕ is indeed unitary, i.e. $u_\phi \in \mathcal{U}(\mathcal{H})$. Notice that the corresponding diffeomorphism in $Diff^+(S^1)$ is given by $\psi(e^{i\theta}) = e^{i\phi^{-1}(\theta)}$. It follows that $\psi \to u_\phi$ defines an anti-representation of $Diff^+(S^1)$ since

$$
\begin{aligned}
\left(u_{\phi_2}(u_{\phi_1}(f))\right)(\theta) &= u_{\phi_1}\left(f(\phi_2(\theta))\right) \cdot |\phi_2'(\theta)|^{\frac{1}{2}} \\
&= f\left(\phi_1(\phi_2(\theta))\right) \cdot |\phi_1'(\phi_2(\theta))\phi_2'(\theta)|^{\frac{1}{2}} \\
&= f\left((\phi_1 \circ \phi_2)(\theta)\right) \cdot |(\phi_1 \circ \phi_2)'(\theta)|^{\frac{1}{2}} \\
&= \left(u_{\phi_1 \circ \phi_2}(f)\right)(\theta)
\end{aligned}
$$

Introduce the splitting, or polarization, $\mathcal{H} = \mathcal{H}_+ \oplus \mathcal{H}_-$, as in section 5.1, when considering the loop group LS^1, and let P be the orthogonal projection onto $\mathcal{H}_-$. It follows that $u_\phi \in \mathcal{U}_2(\mathcal{H}, P)$. That is, we have to prove that $[P, u_\phi]$ is a Hilbert-Schmidt operator.

The action of u_ϕ is represented by the kernel $\delta(\phi(\theta) - \alpha) \cdot \phi'(\theta)^{\frac{1}{2}}$, where δ denotes the Dirac delta function

$$\int_0^{2\pi} \delta(\phi(\theta) - \alpha) \cdot \phi'(\theta)^{\frac{1}{2}} f(\alpha) d\alpha = f(\phi(\theta)) \cdot \phi'(\theta)^{\frac{1}{2}}$$

The polarization operator J on $\mathcal{H} = \mathcal{H}_+ \oplus \mathcal{H}_-$, given by $Jf_\pm = \pm f_\pm$, for $f_\pm \in \mathcal{H}_\pm$, is represented by the singular integral operator

$$(Jf)(\theta) = \frac{1}{2\pi} PV \int_0^{2\pi} K(\theta, \varphi) f(\varphi) d\varphi$$

where PV denotes the principal value of the integral

$$PV \int_0^{2\pi} K(\theta, \varphi) f(\varphi) d\varphi = \lim_{\epsilon \to 0} \left(\int_0^{\theta - \epsilon} + \int_{\theta + \epsilon}^{2\pi} \right) K(\theta, \varphi) f(\varphi) d\varphi$$

and the kernel $K(\theta, \varphi)$ is given by

$$K(\theta, \varphi) = \sum_{k \geq 0} e^{ik(\theta - \varphi)} - \sum_{k < 0} e^{ik(\theta - \varphi)}$$

$$= 1 + i \cdot \cot \frac{1}{2}(\theta - \varphi)$$

which clearly is singular on the diagonal $\theta = \varphi$. Then the kernel $K_{[u_\varphi, J]}$ of the commutator $[u_\varphi, J]$ is given by

$$\int_0^{2\pi} d\alpha \left(\delta(\phi(\theta)) - \alpha) \cdot \phi'(\theta)^{\frac{1}{2}} \cdot K(\alpha, \beta) \right.$$

$$\left. - K(\theta, \alpha) \cdot \delta(\phi(\alpha)) - \beta) \cdot \phi'(\alpha)^{\frac{1}{2}} \right)$$

$$= \phi'(\theta)^{\frac{1}{2}} \cdot K(\phi(\theta), \beta) - K(\theta, \phi^{-1}(\beta)) \cdot (\phi^{-1})'(\beta)^{\frac{1}{2}}$$

since

$$([u_\phi, J]f)(\theta)$$

$$= \int_0^{2\pi} d\alpha \left(\delta(\phi(\theta) - \alpha) \cdot \phi'(\theta)^{\frac{1}{2}} \left(\frac{1}{2\pi} PV \int_0^{2\pi} K(\alpha, \beta) f(\beta) d\beta \right) \right)$$

$$- \frac{1}{2\pi} PV \int_0^{2\pi} d\alpha \left(K(\theta, \alpha) \left(\int_0^{2\pi} \delta(\phi(\alpha) - \beta) \phi'(\alpha) f(\beta) d\beta \right) \right)$$

$$= \frac{1}{2\pi} PV \int_0^{2\pi} d\beta \int_0^{2\pi} d\alpha$$

$$\left(\delta(\phi(\theta) - \alpha) \phi'(\theta)^{\frac{1}{2}} K(\alpha, \beta) - K(\theta, \alpha) \delta(\phi(\alpha) - \beta) \phi'(\alpha)^{\frac{1}{2}} \right) f(\beta)$$

Notice that $K(\theta, \varphi) = 1 + i \cdot \cot \frac{1}{2}(\theta - \varphi)$ is indeed a smooth function in both variables except possibly on the diagonal; in fact

$$K(\theta, \varphi) = \frac{2 \cdot i}{\theta - \varphi} + S(\theta, \varphi)$$

where $S(\theta, \varphi) = S_1(\frac{1}{2}(\theta - \varphi))$ is the smooth function

$$S_1(x) = 1 - \frac{i}{3}x - \frac{i}{45}x^3 - \ldots$$

Evidently $K_{[u_\varphi, J]}$ is smooth off the diagonal, $\theta \neq \varphi$; however, $K_{[u_\varphi, J]}$ is smooth on all of $S^1 \times S^1$. We only need the continuity of $K_{[u_\varphi, J]}$ across the diagonal, to ensure that the kernel $K_{[u_\varphi, J]}$ for $[u_\varphi, J]$ becomes square integrable on $S^1 \times S^1$. By a Taylor expansion to first order in $(\phi^{-1})'(\beta)$, for β sufficiently close to $\phi(\theta)$, it easily follows that

$$\phi'(\theta)^{\frac{1}{2}} \cdot K(\phi(\theta), \beta) - K(\theta, \phi^{-1}(\beta)) \cdot (\phi^{-1})'(\beta)^{\frac{1}{2}}$$

$$= 2i \cdot \left(\frac{\phi'(\theta)^{\frac{1}{2}}}{\phi(\theta) - \beta} - \frac{(\phi^{-1})'(\beta)^{\frac{1}{2}}}{\theta - \phi^{-1}(\beta)} \right)$$

$$\quad + \phi'(\theta)^{\frac{1}{2}} \cdot S(\phi(\theta), \beta) - S(\theta, \phi^{-1}(\beta)) \cdot (\phi^{-1})'(\beta)^{\frac{1}{2}}$$

$$= 2i \cdot \frac{(\phi^{-1})''(\theta)}{(\phi^{-1})'(\theta)}$$

$$\quad + O(\beta - \phi(\theta)) + \phi'(\theta)^{\frac{1}{2}} \cdot S(\phi(\theta), \beta) - S(\theta, \phi^{-1}(\beta)) \cdot (\phi^{-1})'(\beta)^{\frac{1}{2}}$$

which clearly is a bounded continuous function for $|\beta - \phi(\theta)|$ sufficiently small. Here $O(x^n)$ denotes terms in x of order $n \in \mathbf{N}$. This means that the operator represented by the kernel for $[u_\phi, J]$, which naturally is $[u_\phi, J]$, is a Hilbert-Schmidt operator, since its kernel is square integrable on $S^1 \times S^1$. Furthermore, since $[P, u_\phi] = -[u_\phi, \frac{1}{2}(I - J)] = \frac{1}{2}[u_\phi, J]$ it follows that $[P, u_\phi]$ is a Hilbert-Schmidt operator, as claimed.

The associated Lie algebra $diff^+(S^1)$ on $L_2(S^1)$, which is given as the real span of the basis vectors

$$d_k^- = \cos(k \cdot \theta) \cdot d_0 - \frac{1}{2} \cdot k \cdot \sin(k \cdot \theta)$$

for $k \in \mathbf{Z}$ and

$$d_k^+ = \sin(k \cdot \theta) \cdot d_0 + \frac{1}{2} \cdot k \cdot \cos(k \cdot \theta)$$

for $k \in \mathbf{Z} \setminus \{0\}$, where

$$d_0 = \frac{d}{d\theta}$$

in a generalized sense in $L_2(S^1)$. This is in fact a realization of the Lie algebra $Vect(S^1)$ discussed in section 4.2, but with a different choice of basis.

However, it is a bit easier to work with the complexified Lie algebra $diff^+(S^1)_{\mathbf{C}}$ given by the choice of basis vectors

$$d_k = e_k \cdot d_0 + \frac{1}{2}ik \cdot e_k = e_k \left(d_0 + \frac{1}{2}ik \right)$$

where $k \in \mathbf{Z}$ and d_0 are as above. Notice that these (basis vectors as) operators in $L_2(S^1)$ are all unbounded, but all with the same maximal domain given by $\mathcal{D}_{\max} = \{ f \in L_2(S^1) : \sum_{n \in \mathbf{Z}} n^2 \cdot |f_n|^2 < \infty \} = \{ f \in L_2(S^1) : f' \in L_2(S^1) \}$, where f_n, as usual, denotes the n'th Four- ier component of f. Moreover, $d_k^* = -d_{-k}$, since it holds on basis vectors e_m and e_n in $L_2(S^1)$, for arbitrary $m, n \in \mathbf{Z}$

$$
\begin{aligned}
\langle e_m, d_k e_n \rangle &= i\left(n + \frac{1}{2}k\right) \cdot \delta_{m-k-n} = i\left(m - \frac{1}{2}k\right) \cdot \delta_{m-k-n} \\
&= \langle -d_{-k} e_m, e_n \rangle
\end{aligned}
$$

Observe that $d_k^- = \frac{1}{2}(d_k - d_k^*)$ and $d_k^+ = \frac{1}{2i}(d_k + d_k^*)$. Furthermore, we get the following commutation relation

$$[d_m, d_n] = -i(m - n) \cdot d_{m+n}$$

on $\mathcal{D}_{\max}$, which can be verified by a straightforward calculation. This relation is rather close to the corresponding one for the complexification $\eth$ of $Vect(S^1)$, see section 4.2. Finally, we have that the d_k's, $k \in \mathbf{Z}$, fulfil the Jacobi identity (again by a direct calculation). So the algebra spanned by d_k, $k \in \mathbf{Z}$, is indeed a Lie algebra.

Let $u_2(\mathcal{H}, P)_{\mathbf{C}}$ denote the complexification of $u_2(\mathcal{H}, P)$. We will show that d_k belongs to the enlarged Lie algebra of $u_2(\mathcal{H}, P)_{\mathbf{C}}$, allowing unbounded operators, i.e. that $[P, d_k]$ is Hilbert-Schmidt or equivalently $Pd_{-k}(I-P)d_kP$ is of trace-class, for all $k \in \mathbf{Z}$. However, since we later will need the more general fact that

$$Pd_{-k}(I - P)d_m P$$

is of trace-class, for all $k, m \in \mathbf{Z}$, and then especially also for $m = k$, we will show this. Let χ denote the indicator function for $\mathbf{N} \cup \{0\}$, as earlier. Then $d_m P e_n = \chi(-n-1) \cdot i(n+\frac{m}{2})e_{n+m}$ and

$$d_{-k}(I-P)d_m P e_n =$$
$$\chi(-n-1)\chi(n+m) \cdot i^2 \left(n+\frac{m}{2}\right)\left(n+m-\frac{k}{2}\right)e_{n+m-k}$$

giving that

$$\langle e_n, Pd_{-k}(I-P)d_m P e_n \rangle$$
$$= -\left(n+\frac{m}{2}\right)\left(n+m-\frac{k}{2}\right) \cdot \chi(-n-1)\cdot\chi(n+m)\cdot\langle e_n, e_{n+m-k}\rangle$$
$$= -\left(n+\frac{k}{2}\right)^2 \cdot \chi(-n-1) \cdot \chi(n+k) \cdot \delta_{m-k}$$

for arbitrary $k, m, n \in \mathbf{Z}$. Hence

$$\mathrm{Tr}(Pd_{-k}(I-P)d_m P) = \sum_{n \in \mathbf{Z}} \langle e_n, Pd_{-k}(I-P)d_m P e_n \rangle$$
$$= -\sum_{n=-k}^{-1}(n+\frac{k}{2})^2 \cdot \delta_{m-k} \cdot \chi(k)$$
$$= -\delta_{m-k} \cdot \chi(k) \cdot \sum_{n=1}^{k}\left(n-\frac{k}{2}\right)^2$$
$$= -\frac{1}{12} \cdot (k^3+2k) \cdot \delta_{m-k} \cdot \chi(k)$$

which obviously is finite, for all $k, m \in \mathbf{Z}$, it is only non-zero (and then negative) for $m = k > 0$, proving the claim. Hence, $[P, d_k]$ is Hilbert-Schmidt and considerations analogous to those in section 2.3 then imply that $[P, e^{d_k^{\pm}}]$ are Hilbert-Schmidt.

The spin representation of the (hitherto only real) Lie algebra, provides us with a positive energy representation of the Virasoro algebra (see section 4.2). It is of positive energy due to the fact that $H = -idU_P(d_0)$ is non-negative, since $H\Omega = 0$ and on an arbitrary product basis vector $e_{k_1} \wedge \ldots \wedge e_{k_n}$ is

$$H(e_{k_1} \wedge \ldots \wedge e_{k_n}) = \sum_{j=1}^{n}|k_j|(e_{k_1} \wedge \ldots \wedge e_{k_n})$$

because $-i(d_0)_P e_{k_j} = -i d_0 (1 - P) e_{k_j} + i d_0 P e_{k_j} = |k_j| \cdot e_{k_j}$. Since H evidently is an unbounded operator in $\mathcal{F}_\wedge(\mathcal{H})$ we have to specify the domain. Obviously its maximal domains are those vectors $F \in \mathcal{F}_\wedge(\mathcal{H})$, where $\|HF\| < \infty$, we call these vectors finite energy vectors. We choose the algebraic direct sum of all the one-dimensional spaces spanned by the product basis vectors and Ω as the domain $\mathcal{D}(H)$ of H, evidently $\mathcal{D}(H) \subset \mathcal{D}$. As in section 4.3 this give us a natural $\mathbf{N} \cup \{0\}$-grading of $\mathcal{F}_\wedge(\mathcal{H})$

$$\mathcal{F}_\wedge(\mathcal{H}) = \oplus_{m=0}^\infty B_m$$

where B_m is the energy eigenspace corresponding to the eigenvalue $m = \sum_{j=1}^n |k_j|$. Of course, the domain $\mathcal{D}(H)$ is dense in $\mathcal{F}_\wedge(\mathcal{H})$. Moreover, $\mathcal{D}(H)$ is the algebraic direct sum of the B_m's. Notice that H is zero only on $\text{span}\{\Omega, e_0\}$. This construction is very similar to that of the special loop group.

We now complexify the mapping $B \mapsto dU_P(B)$ which until now is only defined for skew-self-adjoint (real linear) elements $B \in u_2(\mathcal{H}, P)$. Since an arbitrary $A \in u_2(\mathcal{H}, P)_\mathbf{C}$ may be decomposed as $A = A^- + iA^+$, with $A^- = \frac{1}{2}(A - A^*)$ and $A^+ = \frac{1}{2i}(A + A^*)$ both in $u_2(\mathcal{H}, P)$, we define the complexification of the mapping $B \mapsto dU_P(B)$ by setting $dU_P(A)_\mathbf{C} = dU_P(A^-) + i \cdot dU_P(A^+)$, where $A = A^- + i \cdot A^+$. Then, we may define the unbounded operator $D_k = dU_P(d_k)_\mathbf{C}$. Of course one have to specify the domains explicitly.

A direct calculation shows

$$D_k\left(e_{k_1} \wedge \ldots \wedge e_{k_n}\right) = \sum_{j=1}^n i\left(k_j + \frac{1}{2}k\right)\text{sign}(k_j + k) \cdot$$
$$\left(e_{k_1} \wedge \ldots \wedge e_{k_{j-1}} \wedge e_{k_{j+k}} \wedge e_{k_{j+1}} \wedge \ldots \wedge e_{k_n}\right)$$

So

$$\|D_k\left(e_{k_1} \wedge \ldots \wedge e_{k_n}\right)\| \leq \left(\sum_{j=1}^n |k_j| + \frac{1}{2}n\,|k|\right) \cdot \|e_{k_1} \wedge \ldots \wedge e_{k_n}\|$$
$$= \left\|\left(H + \frac{1}{2}n\,|k|\right)\left(e_{k_1} \wedge \ldots \wedge e_{k_n}\right)\right\|$$

where the upper bound can be reached, for example for $n = 1$ and $k_1 = k \in \mathbf{N}$. So the maximal domains for all $D_k \circ \mathcal{F}_\wedge(\mathcal{H})$ are equal to each other and equal the set of finite energy vectors, since

$H = -iD_0$. As common domain for all the D_k, $k \in \mathbf{Z}$, we choose $\mathcal{D}(H)$.

Thus $D_k^* = -D_{-k}$ on $\mathcal{D}(H)$ by use of the above complexified mapping and the fact that $d_{-k}^- = d_k^-$ and $d_{-k}^+ = -d_k^+$. Hence, by use of the trace formula derived above, we get

$$
\begin{aligned}
\langle D_k \Omega, D_m \Omega \rangle &= \langle \Omega, -D_{-k} D_m \Omega \rangle \\
&= -\langle \Omega, dU_P(d_{-k})_{\mathbf{C}} dU_P(d_m)_{\mathbf{C}} \Omega \rangle \\
&= -\operatorname{Tr}\left(P d_{-k}(1 - P) d_m P\right) \\
&= \frac{1}{12}\left(k^3 + 2k\right) \delta_{m-k} \cdot \chi(k)
\end{aligned}
$$

where we have used the complexification of formula (2.15) in section 2.3, which can be done by first complexifying in the A-argument,

$$
\begin{aligned}
\langle \Omega, dU_P(A)_{\mathbf{C}} dU_P(B)\Omega \rangle \\
&= \langle \Omega, dU_P(A^-)dU_P(B)\Omega \rangle + i \cdot \langle \Omega, dU_P(A^+)dU_P(B)\Omega \rangle \\
&= \operatorname{Tr}\left(PA^-(I - P)BP\right) + i \cdot \operatorname{Tr}\left(PA^+(I - P)BP\right) \\
&= \operatorname{Tr}\left(P(A^- + iA^+)(I - P)BP\right) \\
&= \operatorname{Tr}\left(PA(I - P)BP\right)
\end{aligned}
$$

and the quite similarly complexify in the B-argument, whence the formula follows. Recall that $\chi(k) = 1$ for $k \in \mathbf{N} \cup \{0\}$, and that $\chi(k) = 0$ for $k \in -\mathbf{N}$, thus $D_k \Omega = 0$ for all $k \leq 0$, since

$$
\|D_k \Omega\|^2 = \frac{1}{12}\left(k^3 + 2k\right) \cdot \chi(k)
$$

Then the associated Lie algebra cocycle becomes

$$
\begin{aligned}
\omega(d_k, d_m)_{\mathbf{C}} &= \operatorname{Tr}(P d_k(I - P)d_m P) - \operatorname{Tr}(P d_m(I - P)d_k P) \\
&= \frac{1}{12}\left(k^3 + 2k\right)\left(\delta_{k+m} \cdot \chi(-k) + \delta_{k+m} \cdot \chi(k)\right) \\
&= \frac{1}{12}\left(k^3 + 2k\right) \delta_{k+m}
\end{aligned}
$$

where we have used the complexification of formula (2.16) in section 2.3, again due to the linearity in the arguments and the earlier derived trace formula. Then the commutation relations of the D_k's are given by

$$
\begin{aligned}
[D_k, D_m] &= [dU_P(d_k)_{\mathbf{C}}, dU_P(d_m)_{\mathbf{C}}] \\
&= dU_P\left([d_k, d_m]\right)_{\mathbf{C}} + \omega(d_k, d_m)_{\mathbf{C}} \cdot I \\
&= -i(k - m)D_{k+m} + \frac{1}{12}\left(k^3 + 2k\right) \delta_{k+m} \cdot I
\end{aligned}
$$

where we have complexified the formula (2.16) in section 2.3 (or formula (2.13) in section 2.4), due to the linearity in the arguments.

It is this commutation relation which shows us that the representation is rather close to be a representation of the Virasoro algebra studied in section 4.2. In fact we can add a particular constant h to the energy operator H such that the cocycle takes the form $\frac{c}{12}k(k^2-1)$, where c is a real constant, and thereby get a representation of the Virasoro algebra. Since the cocycle vanishes for $m \neq -k$, we only need to consider the case $m = -k$, whence the commutation relation may be written as

$$[D_k, D_{-k}] = -i2kD_0 + \frac{1}{12}\left(k^3 + 2k\right) = 2kH + \frac{1}{12}\left(k^3 + 2k\right)$$

Let $H_h = H + h$. Then we may rewrite the commutation relation as

$$\begin{aligned}[D_k, D_{-k}] &= 2k(H+h) - 2kh + \frac{1}{12}\left(k^3 + 2k\right)\\ &= 2kH_h + \frac{1}{12}k\left(k^2 + 2(1-12h)\right)\end{aligned}$$

For $h = \frac{1}{8}$ we get

$$[D_k, D_{-k}] = 2kH_h + \frac{1}{12}k\left(k^2 - 1\right)$$

and the associated cocycle $\omega_{\frac{1}{8}}(d_k, d_m)$ is

$$\omega_{\frac{1}{8}}(d_k, d_m) = \frac{1}{12}k\left(k^2 - 1\right) \cdot \delta_{k+m}$$

and obviously $c = 1$. Hence our representation is labelled by the pair $(h, c) = \left(\frac{1}{8}, 1\right)$, because $\Omega \in \mathcal{D}(H)$.

Before ending this discussion of $Diff^+(S^1)$ as a unitary group, we will analyze the energy operator H, and thereby H_h, by use of the unitary operator S, defined in section 5.1, which raises the charge number of each charge-eigenvectors by one.

It follows that $[H, S] = SQ$, on $\mathcal{D}(H)$. This is trivial on Ω and e_{-1}. For $-1 \notin \{k_1, \ldots, k_n\}$

$$[H, S](e_{k_1} \wedge \ldots \wedge e_{k_n})$$
$$= H\left(e_{k_1+1} \wedge \ldots \wedge e_{k_n+1} \wedge e_0\right) - S\sum_{j=1}^{n}|k_j|e_{k_1} \wedge \ldots \wedge e_{k_n}$$

$$= \left(\sum_{j=1}^{n} |k_j + 1| + 0 - \sum_{j=1}^{n} |k_j| \right) S\left(e_{k_1} \wedge \ldots \wedge e_{k_n} \right)$$

$$= \sum_{j=1}^{n} q_j \cdot S\left(e_{k_1} \wedge \ldots \wedge e_{k_n} \right)$$

$$= SQ\left(e_{k_1} \wedge \ldots \wedge e_{k_n} \right)$$

since each k_j has the same sign as $k_j + 1$ (the sign of 0 is $+$ by definition), so $|k_j + 1| - |k_j| = \pm 1 = q_j$, for $\text{sign}(k_j) = \pm 1$, where q_j is the charge of e_{k_j}. For $k_{n+1} = -1$

$$[H, S](e_{k_1} \wedge \ldots \wedge e_{k_n} \wedge e_{-1})$$

$$= H\left(e_{k_1+1} \wedge \ldots \wedge e_{k_n+1} \right) - \left(\sum_{j=1}^{n} |k_j| + 1 \right) S\left(e_{k_1} \wedge \ldots \wedge e_{k_n} \wedge e_{-1} \right)$$

$$= \left(\sum_{j=1}^{n} |k_j + 1| + 0 - \sum_{j=1}^{n} |k_j| - 1 \right) S\left(e_{k_1} \wedge \ldots \wedge e_{k_n} \wedge e_{-1} \right)$$

$$= \left(\sum_{j=1}^{n} q_j - 1 \right) S\left(e_{k_1} \wedge \ldots \wedge e_{k_n} \wedge e_{-1} \right)$$

$$= SQ\left(e_{k_1} \wedge \ldots \wedge e_{k_n} \wedge e_{-1} \right)$$

since each k_j, $j = 1, \ldots, n$ ($j \neq n+1$) has the same sign as $k_j + 1$, so $|k_j + 1| - |k_j| = q_j$, for $j = 1, \ldots, n$ and $q_{n+1} = -1$. Hence, the commutator formula is proved on product vectors, so the formula holds on $\mathcal{D}(H) \subset \mathcal{D}$. We have earlier (in section 5.1) shown that $[Q, S] = S$ on $\mathcal{D}$. Then, it follows that

$$\left[H - \frac{1}{2}Q(Q - I), S \right] = 0$$

on $\mathcal{D}(H)$, since

$$\left[H - \frac{1}{2}Q(Q - I), S \right] = [H, S] - \frac{1}{2}Q[Q - I, S] - \frac{1}{2}[Q, S](Q - I)$$

$$= SQ - \frac{1}{2}Q(S - 0) - \frac{1}{2}S(Q - I)$$

$$= \frac{1}{2}[S, Q] + \frac{1}{2}S = -\frac{1}{2}S + \frac{1}{2}S = 0$$

by use of the above commutation formulas. Moreover, evidently $[H, Q] = 0$ on Ω and on product basis vectors $e_{k_1} \wedge \ldots \wedge e_{k_n}$, hence on all of $\mathcal{D}(H)$.

So H has the following decomposition as a direct sum

$$H = \oplus_{q \in \mathbf{Z}} H_q$$

where

$$H_q = H|_{\mathcal{H}_q}$$

Recall from section 2.3 that $\mathcal{F}_\wedge(\mathcal{H}) = \oplus_{q \in \mathbf{Z}} \mathcal{H}_q$. Moreover, we put $\Omega_q = S^q \Omega$, where S^q is well-defined for $q \in \mathbf{Z}$, since S is invertible, $S^{-1} = S^*$, and Ω_q belongs to $\mathcal{H}_q$, since $\Omega_q = S^{q-1} e_0 = S^{q-2} e_1 \wedge e_0 = \ldots = e_{q-1} \wedge e_{q-2} \wedge \ldots \wedge e_2 \wedge e_1 \wedge e_0$, for $q > 0$, and $\Omega_q = (S^*)^{-q-1} e_{-1} = (S^*)^{-q-2} e_{-2} \wedge e_{-1} = \ldots = e_q \wedge e_{q+1} \wedge \ldots \wedge e_{-2} \wedge e_{-1}$, for $q < 0$. We call Ω_q a *sector vacuum* (vector).

From the commutation relation above it follows

$$\left(H - \frac{1}{2} Q(Q - I) \right) \Omega_q = S^q \left(H - \frac{1}{2} Q(Q - I) \right) \Omega = 0$$

since $H\Omega = Q\Omega = 0$. Now let $e_{k_1} \wedge \ldots \wedge e_{k_n}$ be an arbitrary product basis vector in $\mathcal{H}_q$, i.e. $Q(e_{k_1} \wedge \ldots \wedge e_{k_n}) = \sum_{j=1}^n q_j \cdot (e_{k_1} \wedge \ldots \wedge e_{k_n})$, where $\sum_{j=1}^n q_j = \sum_{j=1}^n \mathrm{sign}(k_j) = q$. Then

$$H(e_{k_1} \wedge \ldots \wedge e_{k_n}) = \left(\sum_{j=1}^n |k_j| \right) (e_{k_1} \wedge \ldots \wedge e_{k_n})$$

For $q > 0$ we have

$$\sum_{j=1}^n |k_j| \geq \sum_{m=1}^q |k_{j_m}| \geq \sum_{m=1}^{q-1} m$$

$$= \frac{1}{2}(q-1)(q-1+1) = \frac{1}{2}q(q-1)$$

where $k_{j_1}, \ldots, k_{j_m}$ are q arbitrary positive (or zero) indices in increasing order, i.e. $0 \leq k_{j_1} < k_{j_2} < \cdots < k_{j_m}$, it is indeed possible to choose q such indices, since there are exactly q more positive (or zero) than negative indices $k_1, \ldots k_n$ all different from each other (or $e_{k_1} \wedge \ldots \wedge e_{k_n} = 0$).

For $q < 0$ we have

$$\sum_{j=1}^n |k_j| \geq \sum_{m=1}^{-q} |k_{j_m}| \geq \sum_{m=1}^{-q} m$$

$$= \frac{1}{2}(-q)(-q+1) = \frac{1}{2}q(q-1)$$

where $k_{j_1} < k_{j_2} < \cdots < k_{j_{-q}} < 0$ are arbitrarily chosen between the at least $-q$ negative indices $k_1, \ldots k_n$, which is possible since there are $-q$ more negative than positive (or zero) indices, all different from each other.

For $q = 0$

$$\sum_{j=1}^{n} |k_j| \geq 0 = \frac{1}{2}q(q-1)$$

Hence

$$H\left(e_{k_1} \wedge \ldots \wedge e_{k_n}\right) \geq \frac{1}{2}q(q-1)\left(e_{k_1} \wedge \ldots \wedge e_{k_n}\right)$$

$$= \frac{1}{2}Q(Q-I)\left(e_{k_1} \wedge \ldots \wedge e_{k_n}\right)$$

Thus $H \geq \frac{1}{2}Q(Q-I) = \frac{1}{2}q(q-1)$ on product basis vectors in $\mathcal{H}_q$.

Therefore the representation of the Virasoro algebra, given above, restricted to $\mathcal{H}_q$, is caracterized by the label $\left(\frac{1}{8} + \frac{1}{2}q(q-1), 1\right)$, where $\frac{1}{8} + \frac{1}{2}q(q-1)$ is the minimal energy eigenvalue of the new energy operator H_h on $\mathcal{H}_q$ (corresponding to the sector vacuum Ω_q). The earlier mentioned label then corresponds to the sector $\mathcal{H}_0$. Observe that the representations corresponding to q and $-q+1$ give rise to the same label (of course on different sectors), they are therefore unitarily equivalent. Since D_k map $\mathcal{H}_q$ into $\mathcal{H}_q$ and $[H, D_k] = kD_k$, for $k \in \mathbb{Z} \setminus \{0\}$, by the earlier derived commutation relations (or the complexification of formula (2.17)), it follows that $HD_k\Omega_q = D_kH\Omega_q + kD_k\Omega_q = \left(\frac{1}{2}q(q-1) + k\right)D_k\Omega_q$. Hence, the energy of $D_k\Omega_q \in \mathcal{H}_q$ is stricly less than the sector vacuum $\Omega_q \in \mathcal{H}_q$ for any k negative, which is a contradiction, or $D_k\Omega_q = 0$. Whence $D_k\Omega_q = 0$, for any negative $k \in \mathbb{Z}$.

We have analysed the diffeomorphism group $Diff^+(S^1)$, on a Lie algebra level, as a unitary group, by use of the spin representation, treated in section 2.3. Thereby we have constructed positive energy representations of the Virasoro algebra in the charge sectors $\mathcal{H}_q$. These representations all have central charge $c = 1$ and minimal energy $h + \frac{1}{2}q(q-1)$, respectively, where $h = \frac{1}{8}$, corresponding to the energy operator $H_h = H + h$.

There are several ways in which $Diff^+(S^1)$ can act on $\mathcal{H} = L_2(S^1)$, we have treated one realization of $Diff^+(S^1)$ above and will treat another in the next section, namely $Diff^+(S^1)$ as a symplectic group.

5.3 The Diffeomorphism Group $Diff^+(S^1)$ as a Symplectic Group

In section 5.2 we considered $Diff^+(S^1)$ as a subgroup of the restricted unitary group, acting on $\mathcal{H} = L_2(S^1)$, and used the spin representation, constructed in section 2.4 and 2.3. Hence, we obtained a positive energy projective representation on a Lie algebra level, in the anti-symmetric Fock Hilbert space, which turned out to be a sequence of realizations of the Virasoro algebra with central charge $c = 1$. As mentioned earlier this is not the only action of $Diff^+(S^1)$ on the Hilbert space $\mathcal{H} = L_2(S^1)$, in fact, there are several other possible actions (see [Ne, p. 411]).

In this section we will make considerations analogous to those of section 5.2. However, we will consider $Diff^+(S^1)$ as a subgroup of the symplectic group acting on a Hilbert space $\mathcal{H}_0^{1/2}$, and use the metaplectic representation, constructed in section 3.3, to achieve a positive energy projective representation, on a Lie algebra level, in the symmetric Fock Hilbert space. This representation turns out to be a realization of the Virasoro algebra with central charge $c = 1$ and lowest weight $h = 0$.

The following exposition is parallel to the one in section 5.2. Consider the infinite vector space $\mathcal{H}^{1/2}$ of real functions on the unit circle S^1 such that $\sum_{k \in \mathbb{N}} k \cdot |f_k|^2 < \infty$, where $f_k = \langle e_k, f \rangle_{\mathcal{H}}$ is the k'th Fourier component of f with respect to the inner product in $\mathcal{H} = L_2(S^1)$, given in section 5.2. We introduce the semi-inner product on $\mathcal{H}^{1/2}$, as in the case of the spin representation of the special loop group, treated in section 5.1, given in terms of the Fourier components as $\langle f, g \rangle_{1/2} = \frac{1}{2} \sum_{k \in \mathbb{N}} k \cdot (\overline{f}_k g_k + f_k \overline{g}_k)$. Notice that the Fourier components may be complex even though the original functions are real. The only restriction on the Fourier component f_k is $\overline{f}_k = \overline{\langle e_k, f \rangle_{\mathcal{H}}} = \langle e_{-k}, f \rangle_{\mathcal{H}} = f_{-k}$, since f is real. Observe that the semi-inner product given above is in fact a complex inner product on ℓ^2 (for details see section 5.1).

The semi-inner product, given above, is not an inner product since $\mathcal{H}^{1/2}$ has a one-dimensional null space with respect to the semi-norm arising from the semi-inner product $\langle \cdot, \cdot \rangle_{1/2}$. This null space consists of the constant functions $f = f_0$ (where f_0 denote the 0'th Fourier component). Hence, the quotient space $\mathcal{H}_0^{1/2} = \mathcal{H}^{1/2}/\{f :$

$f = f_0 \in \mathbf{R}\}$ is a Hilbert space, with inner product $\langle \cdot, \cdot \rangle_{1/2}$.

First we define the complex unit operator J on $\mathcal{H}_0^{1/2}$ by

$$J\left(\sum_{k \in \mathbf{Z}\setminus\{0\}} f_k e_k\right) = \sum_{k \in \mathbf{N}}(if_k)e_k + \sum_{k \in \mathbf{N}}(-if_{-k})e_{-k}$$

i.e. J multiplies the positive Fourier component f_k, $k \in \mathbf{N}$, by i and the negative Fourier component f_{-k}, $k \in \mathbf{N}$, by $-i$. It follows that $Jf \in \mathcal{H}_0^{1/2}$, for $f \in \mathcal{H}_0^{1/2}$, since

$$
\begin{aligned}
\overline{(Jf)} &= \sum_{k \in \mathbf{Z}\setminus\{0\}} \operatorname{sign}(k)(-i)\overline{f_k}\overline{e_k} = \sum_{k \in \mathbf{Z}\setminus\{0\}} \operatorname{sign}(-k)i f_{-k} e_{-k} \\
&= \sum_{-k \in \mathbf{Z}\setminus\{0\}} \operatorname{sign}(k)i f_k e_k = Jf
\end{aligned}
$$

and

$$\sum_{k \in \mathbf{N}} k\,|(Jf)_k|^2 = \sum_{k \in \mathbf{N}} k\,|\operatorname{sign}(k)i f_k|^2 = \sum_{k \in \mathbf{N}} k\,|f_k|^2 < \infty$$

The above calculations are, of course, equivalent to considering J on $\mathcal{H}^{1/2}$ modulo constant functions.

Obviously $J^2 = -1$, by direct computations, and $J^* = -J$, since

$$
\begin{aligned}
\langle g, Jf \rangle_{1/2} &= \frac{1}{2}\sum_{k \in \mathbf{N}} k\left(\overline{g_k}i f_k + g_k(-i)\overline{f_k}\right) \\
&= \frac{1}{2}\sum_{k \in \mathbf{N}} k\left(\overline{(-Jg)_k} f_k + (-Jg)_k \overline{f_k}\right) \\
&= \langle -Jg, f \rangle_{1/2}
\end{aligned}
$$

Thus, J introduces a complex structure on the set $\mathcal{H}_0^{1/2}$. The complexification $\mathcal{H}_J^{1/2}$ of $\mathcal{H}_0^{1/2}$ is a Hilbert space with respect to the complex inner product given by

$$\langle f, g \rangle_J = \langle f, g \rangle_{1/2} + i\langle Jf, g \rangle_{1/2}$$

Notice that this complex structure is not the usual one. Hence we may write

$$\langle f, g \rangle_J = \frac{1}{2}\sum_{k \in \mathbf{N}} k\overline{f_k}g_k + \frac{1}{2}\sum_{k \in -\mathbf{N}} |k|\,\overline{f_k}g_k$$

$$+ \frac{i}{2} \sum_{k \in N} k \overline{(if_k)} g_k + \frac{i}{2} \sum_{k \in -N} |k| \overline{(-if_k)} g_k$$

$$= \sum_{k \in N} k \overline{f_k} g_k$$

for $f, g \in \mathcal{H}_J^{1/2}$.

Thus, we define the bilinear form $\sigma(\cdot, \cdot)$ on $\mathcal{H}_0^{1/2}$ by

$$\sigma(f, g) = Im \langle f, g \rangle_J$$

Then

$$\sigma(f, g) = \frac{1}{2i} \left(\sum_{k \in N} k \overline{f_k} g_k - \sum_{k \in N} k f_k \overline{g_k} \right) = \frac{i}{2} \sum_{k \in Z \setminus \{0\}} k f_k \overline{g_k}$$

Moreover, we may rewrite $\sigma(f, g)$ as

$$\sigma(f, g) = \frac{i}{2} \sum_{k \in Z \setminus \{0\}} k f_k \overline{g_k} = \frac{1}{2} \left\langle \sum_{n \in Z} f_n e_n, \sum_{k \in Z} ikg_k e_k \right\rangle_{\mathcal{H}}$$

$$= \frac{1}{2} \langle f(\cdot), g'(\cdot) \rangle_{\mathcal{H}} = \frac{1}{4\pi} \int_0^{2\pi} f(\theta) g'(\theta) d\theta$$

for smooth functions $f, g \in \mathcal{H}_0^{1/2}$. Evidently $\sigma(\cdot, \cdot)$ is a non-degenerated symplectic form on $\mathcal{H}_0^{1/2}$.

The natural action of $Diff^+(S^1)$ on $\mathcal{H} = L_2(S^1)$ is given by

$$(s_\phi f)(\theta) = f(\phi(\theta))$$

where ϕ is a smooth real function such that $\phi(\theta + 2\pi) = \phi(\theta) + 2\pi$ and $\phi'(\theta) > 0$ defining the diffeomorphism $\psi \in Diff^+(S^1)$ by $\psi(e^{i\theta}) = e^{i\phi(\theta)}$ (see section 5.2 for further details). In fact, $\psi \to s_\phi$ defines an anti-representation of $Diff^+(S^1)$, since $\psi_2 \circ \psi_1 \to s_{\phi_1 \circ \phi_2}$ and as in section 5.2 one may show that

$$(s_{\phi_2}(s_{\phi_1} f))(\theta) = (s_{(\phi_1 \circ \phi_2)} f)(\theta)$$

Notice that s_ϕ is real linear, with respect to J, and is invertible on $\mathcal{H}_J^{1/2}$, with inverse $s_{\phi^{-1}}$. Moreover, $\sigma(\cdot, \cdot)$ is invariant under s_ϕ,

$$\sigma(s_\phi f, s_\phi g) = \frac{1}{4\pi} \int_0^{2\pi} f(\phi(\theta)) \frac{d}{d\theta}(g(\phi(\theta))) d\theta$$

$$= \frac{1}{4\pi} \int_0^{2\pi} f(\tilde{\theta}) g'(\tilde{\theta}) d\tilde{\theta} = \sigma(f, g)$$

for all $f, g \in \mathcal{H}_0^{1/2}$, where $\tilde{\theta} = \phi(\theta)$. Hence, $s_\phi \in \mathcal{S}p(\mathcal{H}_J^{1/2})$. In fact, s_ϕ is real linear with respect to J, and as usual we split s_ϕ into a complex linear part $(s_\phi)_1 = -\frac{1}{2}J[s_\phi, J]_+$ and a complex anti-linear part $(s_\phi)_2 = \frac{1}{2}J[s_\phi, J]$, which becomes Hilbert-Schmidt. So we only have to prove that $[s_\phi, J]$ is Hilbert-Schmidt, in which case so is $(s_\phi)_2$. The complex structure J only differs from the previous polarization operator, discussed in the beginning of section 5.2 (and there denoted by a J), by the factor i, i.e. $J = iJ_0$, where J_0 is the polarization operator on $\mathcal{H}_J^{1/2}$, that is $J_0 f_\pm = \pm f_\pm$, for $f_\pm \in (\mathcal{H}_J^{1/2})_\pm$ consisting of all vectors in $\mathcal{H}_J^{1/2}$ such that all the negative $(+)$, respectively positive $(-)$, Fourier components vanish (see also the beginning of section 5.1).

Then J and J_0 are singular integral operators with kernel $K(\theta, \varphi)$ $= i - \cot \frac{1}{2}(\theta - \varphi)$ (see the discussion in section 5.2). Since the kernel of s_ϕ is given by $\delta(\phi(\theta) - \varphi)$, the kernel of the commutator $[s_\phi, J]$ is given by

$$\int_0^{2\pi} d\alpha \, (\delta(\phi(\theta) - \alpha)K(\alpha, \beta) - K(\theta, \alpha)\delta(\phi(\alpha) - \beta))$$
$$= \quad K(\phi(\theta), \beta) - K(\theta, \phi^{-1}(\beta))\phi'(\phi^{-1}(\beta))^{-1}$$
$$= \quad K(\phi(\theta), \beta) - K(\theta, \phi^{-1}(\beta))((\phi^{-1})'(\beta))$$

As in the discussion in section 5.2, this kernel is indeed smooth, except possibly on the diagonal $\beta = \phi(\theta)$. However, it is at least continuous at the diagonal. By use of a Taylor expansion, for $|\beta - \phi(\theta)|$ sufficiently small, the prospected singularity in the kernel does not appear, since

$$\frac{1}{\phi(\theta) - \beta} - \frac{(\phi^{-1})'(\beta)}{\theta - \phi^{-1}(\beta)} = \frac{(\phi^{-1})''(\phi(\theta))}{(\phi^{-1})'(\phi(\theta))}$$

by calculations analogous to those in section 5.2. So the kernel is indeed continuous at the diagonal, since ϕ, and then ϕ^{-1}, are strictly monotonous diffeomorphisms. Hence, the kernel is square integrable on $S^1 \times S^1$, which means that the operator it represents, $[s_\phi, J]$, is Hilbert-Schmidt. Therefore $s_\phi \in \mathcal{S}p_2(\mathcal{H}_J^{1/2})$, and we can construct the metaplectic representation of $\mathit{Diff}^+(S^1)$, considered as a symplectic group.

As a basis for the Lie algebra of real vector fields, acting on the Hilbert space $\mathcal{H}_0^{1/2}$, one usually chooses $\cos(k\theta)\frac{d}{d\theta}$ and $\sin(k\theta)\frac{d}{d\theta}$.

However, it is more convenient to use the ordinary complex structures and thereby introduce the basis $d_k = e_k \frac{d}{d\theta}$, $k \in \mathbf{Z}$, where $e_k(\theta) = e^{ik\theta}$. Of course, the operators d_k act in the ordinary complexification $\mathcal{H}_{\mathbf{C}}^{1/2}$ of $\mathcal{H}_0^{1/2}$. Notice that these operators are unbounded, but they have a common maximal domain $\mathcal{D}_{max}$ given by $f \in \mathcal{H}_{\mathbf{C}}^{1/2}$ such that $\|d_k f\|_{\mathcal{H}_{\mathbf{C}}^{1/2}}^2 = \left\|\sum_{n \in \mathbf{Z}} n f_n e_{n+k}\right\|^2 = \sum_{n \in \mathbf{Z}} n^2 |f_n|^2$ is finite. One must be aware that we are now operating with two different complex structures. Later on we will only consider the (ordinary) imaginary part. Of course, there will be no trouble if the (ordinary) complex linear operators commute with J. However, this is not the case for the basis elements d_k, $[d_k, J] \neq 0$, but the commutator is Hilbert-Schmidt and is, in fact, of finite rank. Since $d_k J e_n = -|n| e_{n+k}$ and $J d_k e_n = \text{sign}(n) \cdot \text{sign}(n+k) \cdot (-|n| \cdot e_{n+k})$, $[d_k, J] e_n = |n| \cdot e_{n+k} \cdot (\text{sign}(n) \cdot \text{sign}(n+k) - 1)$ and then

$$
\begin{aligned}
[d_k, J]f &= \sum_{n \in \mathbf{Z} \setminus \{0\}} |n| \cdot f_n \cdot e_{n+k} \cdot (\text{sign}(n) \cdot \text{sign}(n+k) - 1) \\
&= \sum_{n \in \mathbf{Z} \setminus \{0\}} n \cdot f_n \cdot e_{n+k} \cdot (\text{sign}(n+k) - \text{sign}(n)) \\
&= \begin{cases} -2 \cdot \sum_{0<n<-k} n \cdot f_n \cdot e_{n+k} & \text{, for } k \leq 0 \\ 2 \cdot \sum_{-k<n<0} n \cdot f_n \cdot e_{n+k} & \text{, for } k > 0 \end{cases} \\
&= -2 \cdot \sum_{n \in \mathbf{N}_k} |n| \cdot f_n \cdot e_{n+k}
\end{aligned}
$$

where $\mathbf{N}_k = \{1, 2, \ldots, -k-1\}$, for $k < -1$, $\mathbf{N}_k = \{-k+1, -k+2, \ldots, -1\}$, for $k > 1$ and $\mathbf{N}_k = \emptyset$, otherwise.

Observe that $d_k^* = -d_{-k}$ with respect to the ordinary complexification on $\mathcal{D}_{max}$, since on basis vectors e_m and e_n we have

$$
\langle e_m, d_k e_n \rangle_{\mathbf{C}} = \langle -d_{-k} e_m, e_n \rangle_{\mathbf{C}}
$$

where $\langle f, g + i \cdot h \rangle_{\mathbf{C}} = \langle f, g \rangle_{1/2} + i \cdot \langle f, h \rangle_{1/2}$, for $f, g, h \in \mathcal{H}_0^{1/2}$ and $g + i \cdot h \in \mathcal{H}_{\mathbf{C}}^{1/2}$. Hence, $d_k^* = -d_{-k}$. Moreover, on any basis vectors e_k the commutator $[d_m, d_n]$ is well-defined and yields $[d_m, d_n] = -i(m-n)d_{m+n}$ since $[d_m, d_n]e_k = -i \cdot (m-n) \cdot d_{m+n} e_k$. The d_k, $k \in \mathbf{Z}$ do, of course, fulfil the Jacobi identity

$$
[[d_k, d_m], d_n] = [[d_m, d_n], d_k] + [[d_n, d_k], d_m] = 0
$$

So, by linearity, $\{d_k, k \in \mathbf{Z}\}$ span a complex vector space which is a Lie algebra, the ordinary complexification of the Lie algebra corresponding to $Diff^+(S^1)$, considered as a symplectic group.

We now complexify the mapping $A \mapsto dU(A)$ which is defined in section 3.3 for $A \in sp_2(\mathcal{H}_J^{1/2})$ only. First we set

$$d_k^r = \cos(k\theta)\frac{d}{d\theta} = \frac{1}{2}(d_k - d_k^*)$$

and

$$d_k^i = \sin(k\theta)\frac{d}{d\theta} = \frac{1}{2i}(d_k + d_k^*)$$

such that $d_k = d_k^r + i \cdot d_k^i$, where the involution $*$ acts on operators on $\mathcal{H}_\mathbb{C}^{1/2}$. Notice that $d_k^i = -d_{-k}^i = -(d_k^i)^*$ and $d_k^r = d_{-k}^r = -(d_k^r)^*$. Of course their domains are the maximal ones and they are all equal to $\mathcal{D}_{max}$ defined earlier as those $f \in \mathcal{H}_\mathbb{C}^{1/2}$ for which $\sum_{n \in \mathbb{Z}} n^2 |f_n|^2$ is finite, where f_n denotes the n'th Fourier component of f. Both d_k^r and d_k^i are skew-self-adjoint with respect to σ, since

$$
\begin{aligned}
\sigma(d_k^r f, g) &= \frac{1}{4\pi} \int_0^{2\pi} \cos(k\theta) f'(\theta) g'(\theta) d\theta \\
&= \frac{1}{4\pi} \int_0^{2\pi} \cos(k\theta) g'(\theta) f'(\theta) d\theta \\
&= \sigma(d_k^r g, f) = -\sigma(f, d_k^r g)
\end{aligned}
$$

and quite similarly one sees that $\sigma(d_k^i f, g) = -\sigma(f, d_k^i g)$, for all $f, g \in \mathcal{H}_0^{1/2}$.

The complexification of the above mapping $A \mapsto dU(A)$ is then given by

$$dU(d_k)_\mathbb{C} = dU(d_k^r) + i\, dU(d_k^i)$$

We use the abbreviation $D_k = dU(d_k)_\mathbb{C}$. These operators D_k, $k \in \mathbb{Z}$, act in $\mathcal{F}_V(\mathcal{H}_J^{1/2})$. From the general theory, derived in section 3.3, it follows that $dU(d_k^r)$ and $dU(d_k^i)$ are well-defined and essentially skew-self-adjoint on $\mathcal{F}_V(\mathcal{H}_J^{1/2})$. Since $(d_k)_2 = \frac{1}{2}J[d_k, J]$ is a Hilbert-Schmidt operator, so are both $(d_k^r)_2$ and $(d_k^i)_2$, whereby d_k^r and d_k^i both belong to the extended Lie algebra of $sp_2(\mathcal{H}_J^{1/2})$. Therefore, D_k, $k \in \mathbb{Z}$, are well-defined in $\mathcal{F}_V(\mathcal{H}_J^{1/2})$ and fulfil

$$D_k^* = -dU(d_k^r) + i\, dU(d_k^i) = -dU(d_{-k}^r) - i\, dU(d_{-k}^i) = -D_{-k}$$

on the domain $\mathcal{D}$. Furthermore, we put $H = -iD_0 = -i\, dU(d_0)$, then $H^* = i(-D_0) = H$ on $\mathcal{D}$. Since $(d_0)_2 e_n = \frac{1}{2}J[d_0, J]e_n = 0$, by direct calculation, for any $n \in \mathbb{Z}$, we have $(d_0)_2 = 0$ and $(d_0)_1 =$

d_0. Alternatively, one could define P as the orthogonal projection $\frac{1}{2}(I + iJ)$ onto $(\mathcal{H}_J^{1/2})_- = \text{span}\{e_n : n < 0\}$, the $-i$ eigenspace corresponding to J. Then $(d_0)_2 = Pd_0(1-P) + (1-P)d_0P = 0$ by a straightforward caculation, and $d_0 = (d_0)_1 = Pd_0P + (1-P)d_0(1-P)$. Then

$$H(e_{k_1} \vee \ldots \vee e_{k_n})$$
$$= \sum_{j=1}^{n} e_{k_1} \vee \ldots \vee e_{k_{j-1}} \vee k_j e_{k_j} \vee e_{k_{j+1}} \vee \ldots \vee e_{k_n}$$
$$= \left(\sum_{j=1}^{n} k_j\right)(e_{k_1} \vee \ldots \vee e_{k_n})$$

and by the inner product on $\mathcal{F}_\vee(\mathcal{H}_J^{1/2})$ we get

$$\langle e_{k_1} \vee \ldots \vee e_{k_n}, He_{k_1} \vee \ldots \vee e_{k_n}\rangle$$
$$= \left(\sum_{j=1}^{n} k_j\right)\sum_{\sigma \in S_n}\prod_{i=1}^{n}\langle e_{\sigma(k_i)}, e_{k_i}\rangle_J$$
$$= \left(\sum_{j=1}^{n} k_j\right)\sum_{\sigma \in S_n}\prod_{i=1}^{n}(\delta_{\sigma(k_i)-k_i} \cdot k_i \cdot \chi(k_i))$$
$$\geq 0$$

since $\langle e_m, e_n\rangle_J = m \cdot \delta_{m-m}$, for m positive and $\langle e_m, e_n\rangle_J = 0$ for m negative, where $\chi(\cdot)$ denotes the indicator function for $\mathbf{N}$. Notice that it is strictly larger than zero, in the formula, if and only if all the k_j, $j = 1, \ldots, n$ are strictly positive. Hence, H is positive (meaning non-negative). So the representation $d_k \mapsto D_k$ is of positive energy, where $H = -iD_0$ denotes the energy operator in $\mathcal{F}_\vee(\mathcal{H}_J^{1/2})$, with the common domain $\mathcal{D}$.

By the complexification, linearity and the above, we obtain the two-point function

$$\langle D_m\Omega, D_n\Omega\rangle = \langle \Omega, -D_{-m}D_n\Omega\rangle$$
$$= -\langle \Omega, (dU(d_{-m}^r) + i \cdot dU(d_{-m}^i))(dU(d_n^r) + i \cdot dU(d_n^i))\Omega\rangle$$
$$= \frac{1}{2}\text{Tr}_J\left(((d_n^r)_2 + i \cdot (d_n^i)_2)((d_{-m}^r)_2 + i \cdot (d_{-m}^i)_2)\right)$$
$$= \frac{1}{2}\text{Tr}_J\left((d_n)_2(d_{-m})_2\right)$$

where the inner product is that of $\mathcal{F}_v(\mathcal{H}_J^{1/2})$ and formula (3.24) of theorem 17 in section 3.3 has been used. Obeserving that

$$
\begin{aligned}
(d_n)_2 & (d_{-m})_2 e_k \\
&= k(m-k) \cdot (1 - \chi(k)) \cdot \chi(k-m)(1 - \chi(k-m+n))e_{k+n-m} \\
&\quad + k(m-k) \cdot \chi(k) \cdot (1 - \chi(k-m)) \cdot \chi(k-m+n)e_{k+n-m}
\end{aligned}
$$

for k positive and $\chi(\cdot)$ denoting the indicator function for $\mathbf{N}$. Hence

$$
\begin{aligned}
\frac{1}{2}\mathrm{Tr}_J & \left((d_n)_2(d_{-m})_2\right) \\
&= \frac{1}{2} \sum_{k \in \mathbf{Z}\setminus\{0\}} \left\langle \frac{1}{\sqrt{k}}e_k, (d_n)_2(d_{-m})_2 \frac{1}{\sqrt{k}}e_k \right\rangle_J \\
&= \frac{1}{2} \sum_{k \in -\mathbf{N}} k(m-k)\chi(k-m)(1 - \chi(k-m+n))\delta_{n-m}\chi(k) \\
&\quad + \frac{1}{2} \sum_{k \in \mathbf{N}} k(m-k)(1 - \chi(k-m))\chi(k-m+n)\delta_{n-m}\chi(k) \\
&= \frac{1}{2}\delta_{n-m} \sum_{k=1}^{m} k(m-k)\chi(m) \\
&= \frac{1}{12}m(m^2 - 1)\delta_{n-m}\chi(m)
\end{aligned}
$$

where we have explicitly used that $\{\frac{1}{\sqrt{k}}e_k\}_{k \in \mathbf{N}}$ form an orthonormal basis for $\mathcal{H}_J^{1/2}$, since $\langle e_m, e_n \rangle_J = m \cdot \delta_{m-n} \cdot \chi(m)$. Notice that $\|D_m\Omega\| = 0$, for m negative. Therefore the Lie algebra cocycle is

$$
\begin{aligned}
\omega(d_m, d_n)_{\mathbf{C}} &= -\frac{1}{2}\mathrm{Tr}_J([(d_n)_2, (d_m)_2]) \\
&= \frac{1}{12} \cdot m \cdot (m^2 - 1) \cdot \delta_{n+m} \cdot (\chi(m) + \chi(-m)) \\
&= \frac{1}{12} \cdot m \cdot (m^2 - 1) \cdot \delta_{n+m}
\end{aligned}
$$

by complexification of formula (3.28) in section 3.3. The positive energy representation is a so-called level one representation, i.e. $(h, c) = (0, 1)$ where $h = 0$ is the minimal energy and $c = 1$ is the central charge.

We end this section by mentioning that there are other symplectic actions of $Diff^+(S^1)$ than the one considered in this section, for those we refer to [Ne, p. 411].

5.4 The Boson–Fermion Correspondence

In section 5.1 we showed that the mapping $f \to \phi(f)$, $f \in i \cdot slS^1$, provides us with a representation of the canonical commutation relations in the anti-symmetric Fock Hilbert space (see page 175), or equivalently that $e^{if} \to U(e^{if})$ gives a representation of the special loop group SLS^1, which fulfils the Weyl form of the canonical commutation relations (see page 179). As claimed in section 5.1, the first representation mentioned above is unitarily equivalent to the Fock representation $f \to \pi(f)$ on the symmetric Fock space modelled over $\mathcal{H}^{1/2}$, introduced in section 3.1. This remarkable equivalence is well understood by the boson-fermion correspondance. We will not discuss the boson-fermion correspondance in general, but refer to [Mi, p.193-202] and [K-R, p.53-64]; however, we will prove it in this particular case.

Theorem 34 *The "sector energy operator"* $H_\wedge = H - \frac{1}{2}Q(Q - I)$ *is unitarily equivalent to the boson energy operator in each charge sector.*

Proof. We notice that Uhlenbrock, in [Uh], considers a similar correspodence; however, the arguments concerning the mutiplicities are not immediately intelligible, even though the result is correct. Below we give an alternative argument. Since $[d_0, f]g = f' \cdot g$ it follows that $[d_0, f] = f'$ and hence

$$[H, \phi(f)] = -i[dU_p(d_0), \phi(f)] = -i\phi([d_0, f]) = -i\phi(f')$$

by direct calculation, using formula (2.17) and the fact that $\omega_P(d_0, if) = 0$. Hence

$$[H_\wedge, \phi(f)] = [H - \frac{1}{2}Q(Q - I), \phi(f)] = -i\phi(f')$$

by $[Q, \phi(f)] = 0$ (see page 175). This is the expected commutator of $H_\wedge$ with $\phi(f)$.

In section 5.2, page 196 we have already shown that

$$[H_\wedge, S] = [H - \frac{1}{2}Q(Q - I), S] = 0$$

and that the energy operator H may be decomposed as $H = \oplus_{q \in \mathbf{Z}} H_q$, where $H_q = H|_{\mathcal{H}_q}$ and $\mathcal{H}_q$ denotes the q'th charge sector defined in section 2.3, and that $H_\wedge = H - \frac{1}{2}Q(Q - I) \geq 0$ on each $\mathcal{H}_q$ (see page 198). Especially $H_\wedge \Omega_q = H_\wedge S^q \Omega = 0$ (see page 197). We notice that the spectrum of H is $\mathbf{N} \cup \{0\}$ and that the spectrum of Q is $\mathbf{Z}$, whence the spectrum of $H_\wedge = H - \frac{1}{2}Q(Q - I)$ is $\mathbf{N} \cup \{0\}$. Moreover, the boson energy operator in $\mathcal{F}_\vee(\mathcal{H}^{1/2})$, given by the second quantization mapping as $-i \cdot dU(d_0)$, has spectrum $\mathbf{N} \cup \{0\}$.

Below we show that the multiplicities of $H_\wedge$ and of the boson energy operator are the same, and that $H_\wedge$ and the boson energy operator then are unitarily equivalent.

Let us first consider the boson case. In this case any basis product vector $e_{k_1} \vee \cdots \vee e_{k_n} \in \mathcal{F}_\vee(\mathcal{H}^{1/2})$ with $k_1 \geq \ldots \geq k_n > 0$ and energy $\sum_{j=1}^n k_j = m \in \mathbf{N}$ correspond uniquely to a particular partition of $m \in \mathbf{N}$ into a sum of positive integers, i.e. a set $\{k_1, \ldots, k_n\}$ where $k_1 + \cdots + k_n = m$ and $k_1 \geq \cdots \geq k_n > 0$. Moreover, different partitions correspond to orthogonal vectors. The eigenspace $B_m^\vee$ of the boson energy operator corresponding to energy eigenvalue $m \in \mathbf{N}$ is spanned by $e_{k_1} \vee \cdots \vee e_{k_n} \in \mathcal{F}_\vee(\mathcal{H}^{1/2})$, where $\sum_{j=1}^n k_j = m$. Hence, the dimension $\dim(B_m^\vee)$ is exactly the number $p(m)$ of partitions of $m \in \mathbf{N}$ into a sum of positive integers. It can be shown that $p(m) = \frac{d^m}{dx^m}\left(\prod_{n=1}^\infty (1 - x^n)^{-1}\right)|_{x=0}$, but we will not need this result here.

Let us next consider the fermion case. In this case each product vector $e_{j_1} \wedge \cdots \wedge e_{j_n} \in \mathcal{H}_q$ with $j_1 > \cdots > j_n$ and sector-energy $m = \sum_{l=1}^n |j_l| - \frac{1}{2}q(q - 1)$ is uniquely determined by the ordered index-set $(j_1, \ldots, j_n)$, with $j_1 > \cdots > j_n$ and $j_l \in \mathbf{Z}$, $l = 1, \ldots, n$ such that $\mathrm{card}(J_+) - \mathrm{card}(J_-) = q$ and $\sum_{l=1}^n |j_l| - \frac{1}{2}q(q - 1) = m$, where $J_+ = \{j \in \mathbf{N} \cup \{0\} : j \in (j_1, \ldots, j_n)\}$ and $J_- = \{j \in -\mathbf{N} : j \in (j_1, \ldots, j_n)\}$. We call such ordered index-sets *index-tuples*. Hence, we have an isomorphism between the set of orthonormal basis vectors in $\mathcal{H}_q$ and the set of index-tuples (ordered integer-sets) such that the difference between the number of non-negative elements and the number of negative elements is q and such that $\sum_{l=1}^n |j_l| = m + \frac{1}{2}q(q - 1)$. Notice that different index-tuples are mapped into orthonormal basis product vectors.

Define the mapping γ from the set of such index-tuples, defined above, into the set of ordered integer-sequences by $\gamma(j_1, \ldots, j_n) = (i) = (i_{-1}, i_{-2}, \ldots)$, where $i_{-l} = j_l$, if j_l is non-negative and the negative elements $i_{-l} \in (i)$ are the negative integers which do not occur in $(j_1, \ldots, j_n)$. The sequence (i) is ordered in decreasing order $i_{-1} > i_{-2} > \cdots$ and $i_{-l-1} = i_{-l} - 1$ from a certain step $(l \geq -j_n + q + 1)$. We will shortly write this as $\gamma : J_+ \to I_+ = J_+$ and $\gamma : J_- \to I_- = (-\mathbf{N}) \setminus J_-$. We emphasize that $q = \text{card}(I_+) - \text{card}(I_-^c)$, where $I_-^c = (-\mathbf{N}) \setminus I_- = J_-$, and that there exists a $s_0 \in \mathbf{N}$ such that $i_{-s} = q - s$, for $s \geq s_0$. Integer-sequences fulfilling these demands will be called semi-infinite integer-sequences (of charge q). We have

$$\sum_{s=1}^{\infty} (i_{-s} - (q - s)) = \sum_{s=1}^{s_0} i_{-s} - \sum_{s=1}^{s_0} (q - s)$$

$$= \sum_{i \in I_+^{(q-s_0)}} i + \sum_{i \in I_-^{(q-s_0)}} i - \sum_{s=q-1}^{q-s_0} s$$

where $I_\pm^{(q-s_0)} = \{i \in I_\pm : i \geq q - s_0\}$. The rewriting

$$\sum_{i \in I_-^{(q-s_0)}} i = \sum_{i=-1}^{q-s_0} i - \sum_{i \in I_-^c} i$$

then gives

$$\sum_{s=1}^{\infty} (i_{-s} - (q - s)) = \sum_{i \in I_+} i - \sum_{i \in I_-^c} i - \sum_{i=1}^{s_0-q} i + \sum_{s=-(q-1)}^{s_0-q} s$$

$$= \sum_{i \in I_+ \cup I_-^c} |i| - \sum_{s=1}^{|q-1|} s$$

$$= \sum_{j \in J_+ \cup J_-} |j| - \frac{1}{2} q(q - 1)$$

$$= m$$

Notice that γ in fact defines an isomorphism between the set of index-tuples with charge q and sector-energy m and the set of semi-infinite integer-sequences (i) such that $i_{-1} > i_{-2} > \cdots$, $\text{card}(I_+) - \text{card}(I_-^c) = q$, $i_{-s} = q - s$ for s larger than some $s_0 \in \mathbf{N}$ and $\sum_{s=1}^{\infty} (i_{-s} - (q - s)) = m$. Hence, the dimension of the eigenspace $B_m^\wedge(q)$ of the sector-hamiltonian $H_\wedge|_{\mathcal{H}_q}$ corresponding to the energy eigenvalue $m \in \mathbf{N}$ is equal to the number of different ways one can choose semi-infinite integer-sequences fulfilling the above demands.

However, the number of ways one can choose such different semi-
infinite integer-sequences is equal to the number $p(m)$ of partitions
of m into a sum of positive integers (in non-decreasing order). Each
semi-infinite integer-sequence, for fixed $q \in \mathbf{Z}$, can be uniquely written as $(q-1+k_1, q-2+k_2, \ldots, q-n+k_n, q-n-1, q-n-2, \ldots)$
with $k_1 \geq k_2 \geq \cdots \geq k_n > 0$, since $\sum_{s=1}^{\infty} (i_{-s} - (q-s)) = \sum_{s=1}^{n} k_s$.
Moreover, one easily finds that the number of ways to choose a ve-
ctor in $\mathcal{H}_q$ with sector-energy 0 is $p(0) = 1$, namely by the choice
$\Omega_q = S^q \Omega$. Thus

$$\dim \left(B_m^{\wedge}(q) \right) = p(m) = \dim \left(B_m^{\vee} \right)$$

and we may define a mapping $U_{q,m} : B_m^{\vee} \to B_m^{\wedge}(q)$ by first choosing
orthonormal bases for $B_m^{\vee}$ and $B_m^{\wedge}(q)$, respectively, and then let $U_{q,m}$
map the j'th basis vector into the j'th basis vector, $j = 1, \ldots, p(m)$.
By linearity and continuity we extend $U_{q,m}$ to a unitary operator. If
we put

$$U_q = \oplus_{m=0}^{\infty} U_{q,m} : \mathcal{F}_{\vee}(\mathcal{H}^{1/2}) = \oplus_{m=0}^{\infty} B_m^{\vee} \to \mathcal{H}_q = \oplus_{m=0}^{\infty} B_m^{\wedge}(q)$$

then we get a unitary operator mapping $\mathcal{F}_{\vee}(\mathcal{H}^{1/2})$ onto $\mathcal{H}_q$ such
that $U_q H_{\vee} = H_{\wedge}|_{\mathcal{H}_q} U_q$. Whence $H_{\vee}$ and $H_{\wedge}$ are unitary equivalent
in each charge sector $\mathcal{H}_q$ and it follows again, as stated in the end
of section 5.3, that $H_{\wedge}|_{\mathcal{H}_q}$ and $H_{\wedge}|_{\mathcal{H}_{q'}}$ are unitarily equivalent, since
$H_{\wedge}|_{\mathcal{H}_q} = U_q U_{q'}^* H_{\wedge}|_{\mathcal{H}_{q'}} U_{q'} U_q^*$. □

Bibliography

[Ar]   H. Araki, *Bogliubov Automorphisms and Fock Represen-
          tations of the Canonical Anticommutation Relations*, in
          Operator Algebras and Mathematical Physics, edited by
          P.E.T. Jorgensen and P.S Muhly, Comtemporary Mathe-
          matics **62**, 23-141, Providence, Rhode Island 1987.

[A-W]   H. Araki and Wyss, *Representations of Canonical Anti-
          commutation relations*, Helv. Phys. Acta **37**, 136- (1964).

[Arn]   V.I. Arnold, *Mathematical Methods of Classical Mecha-
          nics*, Graduate Texts in Mathematics, Springer-Verlag,
          New York, N.Y. 1978.

[Be]   F. A. Berezin, *The Method of Second Quantization*, Aca-
          demic Press, New York (1966).

[B-P-Z 1] A.A. Belavin, A.M. Polyakov and A.B. Zamolodchikov,
          *Infinite Conformal Symmetry in Two-Dimensional Quant-
          umfield Theory*, Nuclear Phys. **B241**, 333-380 (1984).

[B-P-Z 2] A.A. Belavin, A.M. Polyakov and A.B. Zamolodchikov,
          *Infinite Conformal Symmetry of Critical Fluctuations in
          Two Dimensions*, J. Statist. Phys. **34**, 763-774 (1984).

[B-R 1]   O. Bratteli and D.W. Robinson, *Operator Algebras and
          Quantum Statistical Mechanics I: C*- and W*-Algebras,
          Symmetry Groups, Decomposition of States*, Texts and
          Monographs in Physics, Springer-Verlag, Harrisonburg,
          Virginia 1987.

[B-R 2]   O. Bratteli and D.W. Robinson, *Operator Algebras and
          Quantum Statistical Mechanics II: Equilibrium States Mo-
          dels in Quantum Statistical Mechanics*, Texts and Mono-
          graphs in Physics, Springer-Verlag (1981).

[C-T]   A. Chodos and C.B. Thorn, *Making the Massless String Massive*, Nuclear Phys. **B72**, 509-522 (1974).

[Co]   J.M. Cook, *The mathematics of second quantization*, Trans. Amer. Math. Soc. **74**, 222-245 (1953).

[Dir]   P.A.M. Dirac, *Emission and arbsorbtion of radiation*, Proc. Roy. Soc. London **114**, 248 (1927).

[Dix]   J. Dixmier, *C*-algebras*, North-Holland Publishing Company, Amsterdam, netherlands 1977.

[F-F]   B.L. Feigin and D.B. Fuchs, *Verma Modules over the Virasoro Algebra*, Funct. Anal. Appl. **17**, 241-242 (1983).

[Fo]   V. Fock, *Konfigurationsraum und Zweite Quantelung*, Z. Phys. **75**, 622-647 (1932).

[F-M-S]   D. Friedan, E. Martinee, S. Shenker, *Conformal Invariance, Supersymmetry and String Theory*, Nuclear Physics **B 271**, 93-165 (1986).

[Fr]   K.O. Friedrichs, *Mathematical aspect of the quantum theory of fields*, Comm. Pure Appl. Math. **4**, 161-224 (1951), **5**, 1-56, 349-412 (1952), **6**, 1-72 (1953).

[F-H-J]   J. Fröhlich, G. 't Hooft, A. Jaffe, G. Mack, P.K. Mitter, R. Stora, *New Symmetry Principles in Quantum Field Theory*, NATO ASI series, Series B: Physics vol. 295, Plenum Press, N.Y. (1992).

[G-W 1]   R. Goodman and R. Wallach, *Structure and Unitary Cocycle Representations of Loop Groups and the Group of Diffeomophisms of the Circle*, J. Reine angew. Math. **347**, 69-133 (1984).

[G-W 2]   R. Goodman and R. Wallach, *Projective Unitary Positive-Energy Representations of Diff(S^1)*, J. Funct. Anal. **63**, 299-321 (1985).

[Hov]   L. van Hove, *sur certaines representarions unitaires d'un group infinite de transformations*, Mem. Acad. Roy. Belg. **26**, 68-103 (1951).

[Hu]   J.E. Humphreys, *Introduction to Lie Algebras and Representation Theory*, Graduate Texts in Mathematics, Springer-Verlag, New York, N.Y. 1972.

[J-W]   P. Jordan and E. Wigner, *Pauli's equivalence prohibition*, Z. Phys. **47**, 631 (1928).

[Jø]   P.E.T. Jørgensen, *Operators and representation theory*, Mathematics studies, 147, Notas de Matemática, North-Holland, Elsevier Science Publisher B.V., Amsterdam, Netherlands 1988.

[Ka]   V.G. Kac, *Infinite Dimensional Lie Algebras*, Cambridge University Press, Singapor 1987.

[K-R]   V.G. Kac and A.K. Raina, *Bombay Lectures on Highest Weight Representations of Infinite Dimensional Lie Algebra*, Advanced Series in Mathematical Physics, Vol.2, World Scientific, Singapore 1987.

[Ka]   M. Kaku, *Strings, Conformal Fields, and Topology: An Introduction*, Springer-Verlag, New York, N.Y. 1991.

[La]   E. Langmann, *On Schwinger Terms in (3+1)-dimensions*, to appear (1991). Unpublished.

[Lu 1]   L.E. Lundberg, *Quasi-free "Second Quantization"*, Comm. Math. Phys. **50**, 103-112 (1976).

[Lu 2]   L.E. Lundberg, *Projective Representations of Infinite Dimensional Orthogonal and Symplectic Groups*, KUMI, Denmark (1990). Unpublished.

[L-T]   D. Lust S. Theisen, *Lectures on String Theory*, Lecture Notes in Physics 346, Springer-Verlag, New York, N.Y. 1989.

[Mi]   J. Mickelsson, *Current Algebras and Groups*, Plenum Press, Preprint (1991).

[M-S]   G. Mack and V. Schomerus, *Endomorphisms and Quantum Symmetry of the Conformal Ising Models*, To be published in: Algebraic Theory of Superselection Sectors and Field Theory, Singapore, World Scientific.

[M-N]   F. Murray and J. von Neumann, *On Rings of Operators*,
           Ann. Math.(2) **37**, 116-229 (1936).

[Ne]   Yu. A. Neretin, *Bosons representations of the diffeomorp-
           hism of the circle*, Sovjet Math. Dokl. **28**, 411-414 (1983).

[Ot]   J.T. Ottesen, *Projective Representations of the Loop
           Group and the Boson-Fermion Correspondence* to appear
           in Rep. Math. Phys.

[Pa]   J. Palmer, *Symplectic Groups and the Klein-Gordon Field*,
           Journal of Functional Analysis **27**, 308-336 (1978).

[P-S]   A. Pressley and G. Segal, *Loop Groups*, Clarendon Press,
           Oxford 1986.

[R-S 1]   M. Reed and B. Simon, *Methods of Modern Mathematical
           physics, Vol. I: Functional Analysis*, Academic Press Inc.,
           New York, N.Y. 1980.

[R-S 2]   M. Reed and B. Simon, *Methods of Modern Mathematical
           Physics, Vol. II: Fourier Analysis, Self-Adjointness*, Aca-
           demic Press Inc., Orlando, Florida 1975.

[R-S 3]   M. Reed and B. Simon, *Methods of Modern Mathematical
           Physics, Vol. IV: Analysis of Operators*, Academic Press
           Inc., New York, N.Y. 1978.

[Se,G.]   G. Segal, *Unitary Representations of some Infinite Dimen-
           sional Groups*, Comm. Math. Phys. **80**, 301-342 (1981).

[Se,I.E.] I.E. Segal, *Mathematical Problems of Relativistic Physics*,
           Amer. Math. Soc. Providence, R.I. (1963).

[Sh]   D. Shale, *Linear Symmetries of Free Boson Fields*, Trans.
           Amer. Math. Soc. **103**, 149-167 (1962).

[S-S]   D. Shale and W.F. Stinespring, *Spinor Representations
           of Infinite Orthogonal Groups*, J. of Math and Mech. **14**,
           315-322 (1965).

[V-Z]   A.M Vershik and D.P. Zhenlobenko, *Representation of Lie
           Groups and Related Topics*, Advanced Studies in Contem-
           porary Mathematics, Vol. 7, Gordon and Breach Science
           Publishers, New York, N.Y. 1990.

[Uh]   D.A. Uhlenbrock, *Fermions and associated bosons of one-dimensional model*, Comm. Math. Phys. **4**, 64-76 (1967).

[Vi]   M.A. Virasoro, *Subsidiary conditions and ghosts in dual resonance models*, Phys. Rev. **D1**, 2933-2936 (1970).

[Ya]   K. Yamabana, *On the Algebra Generated by Canonical Commutation Relations*, RIMS, Kyoto Univ.,Preprint.

Index